CONTROLLING THE COST

OF C4I UPGRADES

ON NAVAL SHIPS

JOHN F. SCHANK • CHRISTOPHER G. PERNIN
MARK V. ARENA • CARTER C. PRICE • SUSAN K. WOODWARD

Prepared for the United States Navy

NATIONAL DEFENSE RESEARCH INSTITUTE

The research described in this report was prepared for the United States Navy. The research was conducted in the RAND National Defense Research Institute, a federally funded research and development center sponsored by the Office of the Secretary of Defense, the Joint Staff, the Unified Combatant Commands, the Department of the Navy, the Marine Corps, the defense agencies, and the defense Intelligence Community under Contract W74V8H-06-C-0002.

Library of Congress Cataloging-in-Publication Data is available for this publication.

978-0-8330-4775-5

Cover design by Carol Earnest

Published 2009 by the RAND Corporation
1776 Main Street, P.O. Box 2138, Santa Monica, CA 90407-2138
1200 South Hayes Street, Arlington, VA 22202-5050
4570 Fifth Avenue, Suite 600, Pittsburgh, PA 15213-2665
RAND URL: http://www.rand.org/
To order RAND documents or to obtain additional information, contact
Distribution Services: Telephone: (310) 451-7002;
Fax: (310) 451-6915; Email: order@rand.org

Preface

Command, control, communications, computers, and intelligence (C4I) systems are the lifeblood of naval ships. Various sensors, receivers, computers, and networks gather, process, and disseminate information that both describes the status of the battlefield and the operating condition of the ship and provides sailors the ability to communicate with family and friends.

The C4I capabilities and the architectures that provide those capabilities on naval ships have grown and evolved at a rapid pace over the last few decades, mirroring the technology evolution that has swept the commercial and personal-computing worlds. To keep pace with technology and to take advantage of improvements nurtured in the commercial world, the U.S. Navy extensively uses commercial-off-the-shelf (COTS) hardware and software for its C4I systems; however, COTS products for C4I systems refresh at a fast rate, and the labor cost of just installing new C4I hardware and software on ships amounts to more than $100 million annually.

Recognizing the need to control and reduce the costs of C4I system upgrades, the Program Executive Officer for C4I (PEO C4I) asked the RAND Corporation to examine the factors that influence C4I-upgrade costs and identify what might be done during the design and construction of naval ships to help reduce these costs. This monograph provides the findings and recommendations resulting from RAND's research.

The research was sponsored by PEO C4I and conducted within the Acquisition and Technology Policy Center of the RAND National Defense Research Institute, a federally funded research and development center sponsored by the Office of the Secretary of Defense, the Joint Staff, the Unified Combat Commands, the Department of the Navy, the Marine Corps, the defense agencies, and the defense Intelligence Community.

For more information on RAND's Acquisition and Technology Policy Center, contact the Director, Philip Antón. He can be reached by email at atpc-director@rand.org; by phone at 310-393-0411, extension 7798; or by mail at the RAND Corporation, 1776 Main Street, P.O. Box 2138, Santa Monica, California 90407-2138. More information about RAND is available at www.rand.org.

Contents

Figures

Tables

Summary

The U.S. Navy spends more than $100 million annually on labor to install C4I system upgrades. C4I systems are widely deployed throughout the Navy to gather and process information for decisionmaking and to facilitate communication throughout the ship, among ships and naval bases, and between the ship's company and their families. Expenses associated with installing C4I systems include not only the purchase of new hardware and software but also the labor and material required to install new systems and replace existing systems.

COTS systems allow the Navy to take advantage of investments made in computing technologies by the civilian marketplace, and the Navy has embraced COTS technologies for C4I systems to the extent that the majority of the information technology on naval ships uses commercial hardware, software, and networks. However, it is difficult for the Navy to ensure that naval ships have the most up-to-date systems for computational processing and information sharing because COTS technologies for C4I systems refresh at a rapid pace. COTS systems may change multiple times during the years it takes to build a complex naval warship, and they may also change while in-service ships due for upgrades are unavailable to receive such upgrades because of operational demands.

Research Objectives and Approach

Recognizing the need to control and reduce the costs of C4I system upgrades, PEO C4I asked RAND to examine the factors that influence C4I-upgrade costs and identify what might be done to help reduce those costs. Specifically, we addressed the following questions:

- What actions during the design and construction of naval ships could help reduce the time and cost required to implement future C4I system upgrades?
- What factors contribute to the cost of installing C4I system upgrades on in-service ships, and how well are future C4I system-installation costs estimated?

To understand the challenges and constraints that program offices face when installing C4I systems as they design and build ships and to identify options for reducing C4I installation costs, we interviewed a number of program officials who were involved in the process of building new ships. These programs included *Zumwalt* destroyers, *Ford*-class aircraft carriers, and the next-generation cruiser. We also conducted interviews with program officials associated with ships currently in the active fleet, including *Nimitz*-class aircraft carriers; *Arleigh Burke*–class destroyers; *San Antonio*–class amphibious ships; and *Ohio*-, *Los Angeles*–, and *Virginia*-class submarines. To obtain the shipbuilders' perspective on these challenges and to gather suggestions for reducing costs, we interviewed General Dynamics Electric Boat and Northrop Grumman Newport News.

To analyze cost drivers, variability within costs, and the accuracy of the Navy's cost estimates, we obtained data from PEO C4I, which maintains a database that collects information on C4I-upgrade installations. The data set we obtained contained information on nearly 12,000 system upgrades that took place sometime in 2000–2008. We used techniques from statistics and data mining to look for significant patterns and correlations in the data set, and we tested the significance of potentially significant relationships using statistical analyses that primarily involved correlations, regression, t-tests, and analysis of variance. To assess estimate accuracy, we performed analyses using five error metrics.

Ship-Design and Ship-Construction Initiatives to Reduce C4I-Upgrade Costs

Both the various problems faced when upgrading C4I systems and the actions being taken to overcome those problems can be grouped into three categories: those that arise or apply during new ship construction, those that arise or apply to in-service ships, and those that are common to both new-construction and in-service ships. We address each of these categories below.

C4I-Upgrade Issues Specific to the Design and Construction of New Ships

During the design and construction of new ships, there are four major issues related to the installation costs of C4I systems.

Adopting commercially available systems versus special-built systems. The Navy faces a few disadvantages when it exploits the declining costs and increasing capabilities of the commercial marketplace. COTS equipment is designed for commercial or home use, not for a shipboard operating environment. Peacetime naval operations subject the equipment to salt air and the pitching and rolling of the ship, and even-more-taxing demands occur in wartime environments, which require the equipment to operate successfully after sustaining significant shocks. Military specifications for shock, quieting, and maritime operations can result in the addition of significant modification costs to the otherwise inexpensive commercial equipment. The submarine community has overcome these problems by isolating the equipment from the source of the potential shock. The "rafting" approach it uses for the design and build of the *Virginia* class basically "floats" the equipment above the outside structure of the submarine.

Deciding which C4I systems will be government furnished-equipment (GFE) and which will be contractor-furnished equipment (CFE). There are advantages and disadvantages to both GFE and CFE systems. GFE equipment allows standardization across the fleet, and such equipment is easier to support once a ship enters the fleet. Complete standardization across the fleet of even one C4I system is probably an unattainable goal, however, because of the high refresh rates of commercial technologies. Moreover, ship program managers (PMs) argue that PEO C4I systems are often a generation behind what is available in the open market and that ship designers can provide equal or better capability at lower costs.

CFE systems can leverage the broader commercial marketplace and provide multiple technology options, hopefully at a lower cost. However, CFE systems may be proprietary, and sufficient documentation on the systems may not be available to PEO C4I to support the systems one they enter the fleet. Furthermore, CFE systems can present a problem to PEO C4I once a ship enters the fleet and responsibility for system support transitions to the Navy, which may not have complete knowledge of the systems. Finally, CFE systems may create a unique logistics tail, requiring a separate spare-parts pool and system-unique training for the sailors who operate and maintain the system. The issue of whether PEO C4I should be the preferred provider for new-ship programs warrants further study, especially in terms of the total life-cycle cost impact of choosing either GFE or CFE.

Delivering ships to the fleet with the most-up-to-date systems. One of the biggest problems facing ship PMs during new-ship construction is ensuring that the most-up-to-date C4I technologies are incorporated in a new ship at delivery. C4I technologies can refresh multiple times during the several years it takes to build a ship. Thus, when C4I systems are identified too early in the ship-construction process, they may need to be removed and replaced with the latest upgrade as soon as the ship is delivered to the Navy. On the other hand, ship designers and builders need to lock in both the dimensions and foundations required for the C4I systems and the systems' power and cooling requirements as early as possible in the design and construction process; otherwise, there may not be sufficient space, power, and cooling to support the equipment.

Ship PMs have developed a strategy to overcome the problem of obsolescence in C4I equipment at ship delivery. The strategy—termed *design budget*, *technology insertion*, or *turnkey*, depending on the program—basically provides ship designers and builders the broad specifications of the C4I equipment in terms of space, power, and cooling requirements without identifying the specific equipment that will be installed. Not-to-exceed values for space, power, and cooling are defined, and ship managers must stay within these parameters to ensure that there are no problems installing the equipment when it is finally specified.

Incorporating adequate design margins for weight, power, cooling, and bandwidth into the ship design. The majority of the ships in the Navy fleet have little extra capacity to meet increased demand for weight, power, cooling, and bandwidth because the initial design margins are typically consumed very early in a ship's life. This lack of extra capacity results in difficulties in finding ship services to support the new equipment and in additional costs when upgrading C4I capabilities, especially when adding new capabilities to a ship. Workarounds, including the addition of new power and cooling plants, are often needed to provide the additional ship services required to accommodate C4I upgrades. Existing systems may have to be downgraded, or ship operations may need to be constrained (e.g., by not operating certain systems while others are being operated). Some ship classes have become so constrained that a new system or capability cannot be added without an existing system being removed.

The issue of adequate design margins is being addressed in the design of new classes of ships. For example, the design for the new *Ford* class of aircraft carriers includes both extra cabling in various spaces and extra bandwidth. The design also features a zonal electricity grid that allows power to be directed throughout the ship where and when it is needed.

C4I-Upgrade Issues for In-Service Ships

Many of the issues that arise during ship design and construction also arise when upgrading C4I systems for ships that are in the operational fleet. There are, however, two problems that are unique to in-service ships.

Various ship configurations make the planning of upgrades difficult. The cost of specific upgrades to C4I systems could be reduced if the installation details were the same across all ships in a given class. This uniformity would permit both the creation of just one set of design drawings for the upgrade and the implementation of a repetitive approach to installing the upgrades. Unfortunately, the C4I systems of various ships in a given class feature different configurations, and the areas where C4I systems are installed may be differently laid out. Nonstandard configurations across ships in a single class mean that each installation requires a check of the ship's configuration, the development of a unique set of design drawings, and an almost-unique process for installing the upgrade. These requirements contribute to increased cost and time to accomplish the upgrade. The biggest contributor to different configurations is the high refresh rate of the technology used in the C4I systems. For example, the C4I configurations in DDG-51–class ships currently in construction will differ from those in DDG-51–class ships delivered only a few years ago. The submarine community appears to do the best job of maintaining similar C4I configurations across all of the submarines within a class and even across various classes. Part of this success is due to adapting an open-architecture design philosophy and business model for C4I and combat systems.

Navigating the ship maintenance (SHIPMAIN) process. The SHIPMAIN process was initiated in November 2002 to identify and eliminate redundancies in maintenance processes so that the right maintenance is done at the right time, at the right place, and at the right cost. It also seeks, through a common planning process for ship maintenance and modernization, to maintain configuration control of the various changes made to ship systems and equipment over the life of a ship. SHIPMAIN has been successful in reducing the churn in maintenance and modernization planning and in reducing maintenance costs while providing ready ships to the fleet. It has also been successful in establishing and sustaining configuration control across the ships in a given class for hull, mechanical, and electrical equipment and systems. However, because C4I technologies can change one or two times during the three-year planning and implementation period, SHIPMAIN is typically viewed as too difficult and too time-consuming to implement during C4I upgrades. The process—and the time required to approve and implement changes—is the same regardless of the magnitude of the change. For example, software upgrades go through the same SHIPMAIN process as major hardware changes.

Software upgrades are also more difficult than hardware changes to certify in the SHIPMAIN process because it is difficult to specify the impact of a software change from an operational perspective.

In interviews, various organizations noted that, with waivers, an alteration can be approved within 90 days under SHIPMAIN. However, the approval process itself takes time and resources. Because quick approvals are typically required for C4I upgrades, this shorter alternative should be streamlined and made easier to navigate.

General C4I-Upgrade Issues

A few factors influence C4I installation costs during both new-ship construction and when upgrading in-service ships: the need to integrate and test the new systems installed in the ship, the amount of "hot work" (as welding and the installation of foundations and structures are known) and changes to ship services (i.e., space, power, and cooling) required, and the need to integrate the antennas on the topside of the ship. There are factors specific to ship design and construction and factors specific to in-service ships.

Integrating and testing the new systems installed in the ship. The proliferation of C4I systems on naval ships has complicated the process of integrating the various systems into an overall ship C4I architecture and testing the overall C4I package to ensure that all functions are working correctly. How systems integration will occur is a key decision in the C4I design process. The federated approach decentralizes the hardware and software functions of the various C4I systems while allowing all systems to share data and information through a common network. Under this decentralized approach, the hardware or software problems of one C4I system, or of an upgrade to a system, can be isolated and addressed without affecting the performance of other C4I systems. One downside of using a federated approach is that doing so involves some level of redundancy in C4I hardware and software, which results in increased costs and increased demands on ship services.

An integrated architecture reduces hardware and software redundancies by using both a shared network and shared computing hardware and system software. C4I functions in integrated systems are typically supplied by software programs that use common system-hardware processing and the common network to share data and information. The disadvantages of integrated systems are (1) the potential loss of all C4I functions when a problem with the common hardware or system software arises and (2) the fact that there are few suppliers capable of delivering the more-complex hardware and software systems involved in this type of architecture.

Regardless of the overall architecture chosen for C4I systems, a consolidated testing plan is needed to ensure that all C4I systems are working correctly in both a stand-alone mode and as part of the overall C4I architecture. A consolidated testing plan should reduce redundant activities, thus saving money, and allow testing to occur later in the build process. However, a consolidated testing plan typically requires that a facility perform the testing of the overall C4I system, and constructing and operating such a facility entail monetary, schedule, and opportunity costs. The *Virginia* program has successfully implemented a consolidated testing plan in which the electronic components of the C4I weapon systems are assembled and tested at the Command and Control System Module Off-Hull Assembly and Test Site facility at Electric Boat before they are inserted into the submarine.

The need for both hot work and changes to ship services drives the costs of C4I upgrades. If adding a new capability to or upgrading an existing capability on a ship were as easily accomplished as upgrades to home computers or audio-visual systems, C4I-upgrade costs would not be an issue for the Navy. However, the complexity of integrating C4I systems, the limited supply of ship services, and the density of modern ships require a significant amount of labor to remove and replace equipment. The Navy is employing several initiatives to reduce the required amount of hot work and changes to ship services. For example, the CVN-21 aircraft-carrier program[1] is incorporating a flexible infrastructure into various C4I spaces on the *Ford* class, and changes since the last design include a raised deck with ventilation running underneath and movable vents on the deck to direct the cooling where it is needed. New-ship designs are also incorporat-

---

[1]   A CVN is an aircraft carrier, nuclear.

ing features to allow for easier access to C4I systems during upgrade installations. For example, wider passageways and less-dense placement of C4I equipment allow for the removal and installation of new equipment without affecting existing equipment.

Integrating antennas for the various C4I systems on the topside of the ship is often more difficult than integrating C4I systems within the ship structure. Each antenna must have a clear field of view to receive signals, and it cannot cause electronic interference with other antennas. Topside space, from both the horizontal and vertical perspectives, is limited and must be carefully managed. There are efforts under way to consolidate various antennas, but these efforts have not yet led to an acceptable solution. The submarine community has made progress with the topside integration problem by adopting a universal modular mast that allows antennas to be changed more easily. Table S.1 summarizes these various upgrade problems and displays options aimed at lessening their impact during C4I upgrades. As the table shows, single management decisions and design options are typically aimed at solving multiple upgrade problems.

Factors That Influence C4I Installation Costs and the Accuracy of Estimates

Through our analyses of historical data on the cost of installation labor, we sought to better understand the factors that influenced the labor costs of installing certain prior C4I upgrades, the variability within those costs, the extent to which cost improvement occurred, and the accuracy of cost estimates. Table S.2 shows the labor costs of the six types of upgrades that were the focus of our analyses.

Our analyses found inconsistent trends across the types of installations. Although there were significant differences within an installation type (e.g., installation on one coast was more expensive than on the other), there were no consistent trends associated with factors, such as age and size of ship, that would be useful in adjusting future estimates. This variability is a reflection of both the quality of the data and the diversity of the installations, which ranged from

Table S.1
Actions Taken to Solve Various C4I-Upgrade Problems

Action Taken	Cost of Hot Work	Ship Service Modifications	Military Specifications	Systems Integration or Testing	High Tech Refresh Rate	Requirements Growth
Create a flexible infrastructure	X	X			X	X
Employ standard racks or enclosures	X	X			X	X
Employ modular isolation			X			
Improve access to C4I-system spaces	X	X			X	
Use the design-budget process during construction					X	X
Incorporate adequate design margins					X	X
Decide on federated vs. integrated systems			X	X		
Employ consolidated testing				X	X	
Use COTS, SOA, and/or OA					X	X

Table S.2
Labor Costs of Analyzed Upgrades

Upgrade Type	Number of Installations Examined	Labor Cost (FY09$)		
		25th Percentile	Average	75th Percentile
GCCS-M GENSER 4.X (V) 1–4	31	158,121	283,651	459,345
ISNS Embarkable Drops	34	55,016	186,600	188,751
ISNS LAN GIG-E	54	1,029,918	1,577,444	1,624,571
ISNS COMPOSE 3.0 Software	82	27,947	43,303	51,617
EBEMs for WSC-6 Variants	48	32,212	42,929	50,682
SSEE Increment E	24	545,300	734,060	918,195

software upgrades to major system replacements. For example, ship size and age frequently had a significant impact on the cost of installation related to Ships Signal Exploitation Equipment (SSEE), super-high frequency radios, the Integrated Shipboard Networking System (ISNS), and the Global Command and Control System–Maritime. In some cases, the costs varied significantly by installation location. Also, for some specific installations, the variability in actual labor cost, even within a ship class, was quite high (i.e., the high-to-low value was different by an order of magnitude). Early adopters of the Consolidated Afloat Networks and Enterprise Services system tended to exhibit higher installation costs than did similar ships that were not selected as early adopters.

There was some evidence of cost improvement but it was weak and inconsistent across upgrade types. In many industrial and manufacturing situations, costs decrease as an activity is accomplished more frequently. Our analyses of the different upgrades suggest that there was a decrease in installation costs associated with both the Enhanced Bandwidth-Efficient Modem for WSC-6 Variants and the ISNS Embarkable Drops upgrades. For these upgrades, costs decreased as successive installations were accomplished. However, there was negative learning (i.e., an increase in costs for successive upgrades) for the ISNS Common PC Operating System Environment and SSEE Increment E upgrades. This finding is troublesome because it suggests that teams required more hours to complete each successive installation of these upgrades—a trend exactly the opposite of what we expected.

We used several metrics to assess the accuracy of the estimates of C4I-upgrade costs. The mean bias across all the installations in the database was over $7,000, suggesting that installation costs were typically overestimated. Overestimating was a particular problem for installations on larger ships, such as aircraft carriers and amphibious ships. Also, overestimates were especially large for installations that had low actual costs. Furthermore, we found that cost estimations tended to overestimate the cost of upgrades. Additionally, the relative error was quite high, particularly for aircraft carriers, nuclear; guided missile frigates; and attack submarines, nuclear. Many of the factors that were found to influence cost variability, such as hull type, ship age, and installation coast, also affected estimation accuracy.

Recommendations

Several recommendations flow from these findings. Although our first few recommendations are not new, they are not being considered consistently across ship types and classes. From the perspective of designing and building naval ships to facilitate C4I upgrades, the Navy should

- ensure that adequate design margins for power, cooling, and space are incorporated into the design of a ship and that adequate margins are sustained during the operational life of a ship
- include adequate access paths when designing a ship, especially a surface combatant
- use standard racks and fixtures and flexible infrastructures to help reduce the amount of hot work required to remove old fixtures and install new ones when upgrading C4I systems

- conduct additional analyses comparing the advantages and disadvantages of using (1) GFE versus CFE, (2) federated versus integrated systems, and (3) service-oriented architectures versus open architectures
- include in the Space and Naval Warfare Systems Center/PEO Integrated Data Environment and Repository database both information about the facility where an installation was performed (i.e., private shipyard, public shipyard, or operating base) and an identifier for the organization or team that accomplished the installation
- develop after-action reports for each installation or, if that is not feasible, for at least those installations during which actual costs significantly differed from estimated costs.

Acknowledgments

This research was initiated and supported by Chris Miller, the PEO C4I. We are very appreciative of the help provided by CAPT Joe Beel, a deputy PM in the PEO C4I organization, who facilitated our interactions with PEO C4I staff. Numerous people on the PEO C41 staff— too many to mention here—provided valuable information about and insights into the issues and problems surrounding C4I upgrades on naval ships. We thank them all. Of special note are Ruth Youngs Lew and Leo Martinez, who provided the data used in our analyses and helped explain the intricacies of the Space and Naval Warfare Systems Center/PEO Integrated Data Environment and Repository database.

Debbie Peetz of RAND provided her typical excellent overall support to the research. Isaac Porche and Kevin Brancato, also of RAND, reviewed an earlier version of the manuscript and provided numerous constructive comments that strengthened the monograph.

Of course, any errors of omission or commission in the document are the sole responsibility of the authors.

Abbreviations

ACAT	acquisition category
AEHF	advanced extremely high frequency
ANOVA	analysis of variance
C2	command and control
C2P	C2 Processor
C4I	command, control, communications, computers, and intelligence
CANES	Consolidated Afloat Networks and Enterprise Services
CDLMS	Common Data Link Management System
CEC DDS	Cooperative Engagement Capability Data Distribution System
CFE	contractor-furnished equipment
CG	guided missile cruiser
CLIP	Common Link Integration Processing
CNO	Chief of Naval Operations
COMPOSE	Common PC Operating System Environment
COP	common operational picture

COTS	commercial off-the-shelf
CVN	aircraft carrier, nuclear.
DDG	guided missile destroyer
DJC2	Deployable Joint Command and Control
DSCS	Defense Satellite Communications System
EBEM	enhanced bandwidth-efficient modem
EDSS	Expeditionary Decision Support System
EHF	extremely high frequency
FFG	guided missile frigate
FY	fiscal year
GBS	Global Broadcasting System
GCCS	Global Command and Control System
GCCS-J	GCCS–Joint
GCCS-M	GCCS–Maritime
GENSER	General Services
GFE	government-furnished equipment
GIG	Global Information Grid
GIG-E	GIG–Electronic
HF	high frequency
Inmarsat	International Maritime Satellite
INTELSAT	Intelligence Satellite
ISNS	Integrated Shipboard Networking System
JEM	Joint Effects Model
JOEF	Joint Operational Effects Federation

JPEN	Joint Protection Enterprise Network
JSIMS-M	Joint Simulation System–Maritime
JSS	Joint Interface Control Officer Support System
JTIDS	Joint Tactical Information Distribution System
JTRS WNW	Joint Tactical Radio System–Wideband Networking Waveform
JWARN	Joint Warning and Reporting Network
LAMPS	Light Airborne Multipurpose System
LAN	local area network
LCC	amphibious command ship
LHA	amphibious assault ship
LHD	amphibious assault ship (dock)
Ln	natural logarithm
LOS	line of sight
LPD	amphibious transport dock
LSD	landing ship dock
MIDS	Multifunctional Information Distribution System
MIDS-LVT	MIDS–Low Volume Terminal
MOS	MIDS on Ship
MUOS	Mobile User Objective System
NCES	Net-Centric Enterprise Services
NESP	Navy Extremely High Frequency SATCOM Program
NMT	Navy Multiband Terminal

NTCSS	Naval Tactical Command Support System
OA	open architecture
PC	personal computer
PEO	program executive office/program executive officer
PEO C4I	PEO for C4I
PM	program manager
PMW	Program Manager, Warfare
POR	program of record
PSA	postshakedown availability
R^2	coefficient of determination
RCOH	refueling complex overhaul
RDC	Rapid Deployment Capability
SATCOM	satellite communications
SCI	sensitive compartmented information
SHF	super-high frequency
SHIPMAIN	ship maintenance
SOA	service-oriented architecture
SSBN	fleet ballistic missile submarine
SSEE	Ships Signal Exploitation Equipment
SSN	attack submarine, nuclear
TADIL	Tactical Digital Information Link
TBMCS	Theater Battle Management Core System
TIDS	Tactical Integrated Digital System
TSAT	Transformational SATCOM System

UFO	UHF Follow-On
UHF	ultra-high frequency
VHF	very high frequency
WGS	Wideband Global Satellites
WOO	window of opportunity

CHAPTER ONE

Introduction

The Problem: Controlling and Reducing the Costs of C4I Upgrades

Command, control, communications, computers, and intelligence (C4I) systems are some of the most important components on naval ships. Through a variety of sensors, receivers, computers, networks, and software, diverse information is received, processed, and disseminated both throughout the ship and to other ships in the battlegroup. C4I systems provide continuous status information about the operational condition of the ship and support net-centric–warfare concepts by reporting on the environment in which the ship and other ships in the battlegroup are operating. C4I systems also support improved quality of life for the ship's company by providing sailors the ability to communicate with family and friends.

The commercial world has fueled the growth and evolution in the computing and information technology that is the backbone of the U.S. Navy's C4I systems. To take advantage of both the economies offered by commercial products and the most-up-to-date technologies available, the Navy uses commercial-off-the-shelf (COTS) products for much of the C4I hardware and software on naval ships. Adopting commercial standards and practices has allowed the Navy to both improve existing C4I capabilities and add new computing capabilities to naval ships. However, rapid advances in information technology present challenges to Navy planners.

Like most consumers, the Navy wants the most-up-to-date technologies and equipment. However, getting these items aboard is difficult to accomplish when building new ships because some aspects of the information technology may evolve and change during the several years it takes to build a complex naval warship. The Navy must balance the shipbuilders' desire to specify design parameters early in the construction process with its own desire to have the most-up-to-date C4I systems installed when the ship enters the operational fleet.

The problem does not become easier once a ship enters the fleet. Operational demands limit when a ship is available for either the installation of new C4I systems or the upgrade of existing ones. A ship may enter a Chief of Naval Operations (CNO) availability[1] every two to three years, and the repair and modernization work accomplished during availabilities, as well as the funding available to do it, is often defined a year or more in advance. With technologies changing rapidly and a limited ability to install the most-recent technologies, it is difficult or even impossible to standardize C4I technologies across the fleet or even across a given class of ships.

Installing the most-up-to-date technologies while trying to standardize C4I systems across the fleet is further complicated by the constrained budgets available for fleet modernization. A given C4I installation package for a ship can cost more than $100,000. Currently, the Navy spends more than $100 million annually just on labor to install C4I upgrades.

Research Objectives and Approach

Recognizing the difficulties associated with controlling the costs of C4I upgrades, the Program Executive Officer (PEO) for C4I (PEO C4I) asked RAND to determine what could be done during the design and

---

[1]   A CNO availability is a scheduled maintenance period in a private or public shipyard. All naval ships are on readiness cycles that involve training, deployment, and maintenance. The length of these cycles, including the duration of the maintenance availability, varies for different classes of ships. For example, the current *Nimitz*-class aircraft carrier cycle is 32 months, which includes a six-month nondocking availability or a 10-month docking availability. For more information on the cycles of aircraft carriers, see Yardley et al., 2008.

construction of naval ships to reduce the costs of future C4I upgrades. Specifically, we addressed the following questions:

- What actions during the design and construction of naval ships could help reduce the time and cost required to implement future C4I system upgrades?
- What factors contribute to the cost of installing C4I system upgrades on in-service ships, and how well are future C4I system installation costs estimated?

We pursued two paths to address the research questions. First, we interviewed a number of program officials who were involved in the process of building new ships. We wanted to understand the challenges and constraints they faced in designing and building the ships to facilitate the introduction of new C4I capabilities. We also wanted to understand what, if anything, they were doing to make future C4I upgrades easier and, hopefully, less costly.

Second, we interviewed a number of officials involved in ship programs that were still in either the design phase or the early construction stage. In these interviews, we were most interested in learning what methods the officials were considering during the design of their C4I systems, especially in regard to facilitating future C4I upgrades.

We conducted interviews in several ship programs. In the active fleet, we interviewed officials associated with *Nimitz*-class aircraft carriers; *Arleigh Burke*–class destroyers; *San Antonio*–class amphibious ships; and *Ohio*, *Los Angeles*, and *Virginia*-class submarines. In the design or early construction phases, we interviewed officials associated with *Zumwalt* destroyers, *Ford*-class aircraft carriers, and the next-generation cruiser. We rounded out both sets of interviews by talking to two shipbuilders—General Dynamics Electric Boat and Northrop Grumman Newport News—to understand their perspectives and to obtain suggestions for reducing upgrade costs.

To analyze cost drivers, variability within costs, and the accuracy of the Navy's cost estimates, we obtained data from PEO C4I, which maintains a database that collects information on C4I-upgrade installations. The data set we obtained contained information on nearly

12,000 system upgrades that took place sometime in 2000–2008. We used techniques from statistics and data mining to look for significant patterns and correlations in the data set, and we tested the significance of potentially significant relationships using statistical analyses that primarily involved correlations, regression, t-tests, and analysis of variance. To assess estimate accuracy, we performed analyses using five error metrics.

Organization of the Report

Chapter Two provides background information on Navy C4I systems, describing related guidance, organization, trends, and specific systems. Chapter Three summarizes a number of the issues the Navy and ship-builders face when considering C4I upgrades during the ship-design and ship-construction processes and after ships have entered the operational fleet. It also describes various actions the Navy and shipbuilders are taking to help facilitate C4I upgrades and control their costs. Chapter Four describes our analyses of the labor costs associated with installing C4I upgrades and assesses both variability within these costs and the accuracy of cost estimates. Chapter Five offers conclusions and recommendations. Appendix A lists programs currently managed by PEO C4I, and Appendix B describes the upgrade types we analyzed.

CHAPTER TWO

Background on Navy C4I Systems

C4I systems constitute the hardware, software, and processes that enable decisionmakers to generate information and use it to command and control (C2) their forces. In this chapter, we describe an overarching concept that has helped codify the Navy's definition of a *networked force*. We also describe several organizations that have responsibility for C4I programs, emphasizing PEO C4I. Finally, we examine the Navy's C4I capabilities and related systems.

Sea Power 21 and FORCEnet

In October 2002, the CNO introduced the concept of Sea Power 21, describing the Navy through four capabilities: Sea Strike, Sea Shield, and Sea Basing, which constitute the offensive, defensive, and operational aspects of the force, respectively, and FORCEnet,[1] which is the Navy and Marine Corps' vision for future information-centric operations. FORCEnet is not a system or a materiel solution. It is a construct approved by the CNO and by the Commandant of the Marine Corps in 2005 for integrating the Navy's people, sensors, networks, C2, and weapons among themselves and with outside partners. The Director of Net-Centric Warfare has responsibility for oversight of FORCEnet.[2]

---

[1] For additional information on FORCEnet, see Naval Network Warfare Command, undated.

[2] See Deputy Chief of Naval Operations, Warfare Requirements and Programs (N6/N7), 2005a.

The Army and the Air Force have similar paradigms, known as Land-WarNet and C2 Constellation, respectively, with which the Navy has aligned itself. FORCEnet has support from the highest authorities within the Navy, and PEOs have integrated FORCEnet tenets into their choice of C4I solutions for the Navy.

Responsibility for C4I

PEO C4I acquires, fields, and supports C4I systems across Navy, joint, and coalition forces.[3] Overall, its goals are to adapt C4I technologies to the rapidly expanding capabilities available in the commercial marketplace and do so while being responsive to the requirements of the fleet, reducing overall system and life-cycle costs, and aligning with the joint community. Among its specific objectives is trying to decrease the installation costs of C4I systems by 10 percent and make concomitant reductions in the variability of cost estimates.[4] PEO C4I averages more than 2,700 installations of C4I equipment per year.[5]

PEO C4I has program management responsibility for approximately 150 C4I systems across ten program offices. About one-third of the programs it manages are programs of record (PORs), some of which are described in Appendix A; the others are acquisition projects that either will merge with existing PORs or with other non-PORs or will become their own PORs. Major Defense Acquisition Programs for which PEO C4I has responsibility include

- Deployable Joint Command and Control
- Global Command and Control System (GCCS)–Maritime (GCCS-M)
- Navy Extremely High Frequency Satellite Communications Program

---

[3] PEO C4I, undated-a.

[4] U.S. Navy Program Executive Officer, Command, Control, Communications, Computers, and Intelligence Team, 2007b, p. 12.

[5] PEO C4I, 2008.

- Navy Multiband Terminal
- Naval Tactical Command Support System.

The PEO is involved in joint programs, including the Joint Tactical Radio System–Airborne, Maritime, and Fixed Stations; Common Link Integration Processing; and Deployable Joint Command and Control.

Other Navy organizations also manage C4I systems. For example, the PEO for Integrated Warfare Systems is responsible for developing C2 systems for more than 100 programs, and the PEO for Warfare develops C2 systems for strike weapons and unmanned aviation platforms.

Navy C4I Systems

The Navy employs a large number and wide variety of C4I systems, including both software and hardware, across the fleet. An abstracted representation of the three layers of C4I systems that might be aboard a Navy vessel is shown in Figure 2.1:

- At the bottom is the network layer, which includes the switches, routers, and collection devices that allow a ship to move information into and out of the ship and among its many applications. This layer includes the hull, mechanical, and electrical functions and networks that run the ship and transport information.
- In the middle is the service layer, which includes the larger servers that provide email, storage, and common software used aboard a ship.
- The top layer is the application layer, where sailors interact with applications through the local area network (LAN) and individual computers and workstations. In this layer, computers allow individual sailors to interact with software packages, such as the GCCS, to access information that has been transported from external sensors and now resides on the ship's servers.

Figure 2.1
Abstracted Representation of C4I Systems

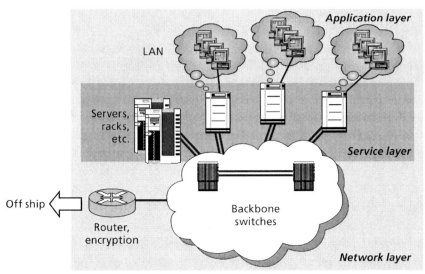

SOURCE: Adapted from Program Executive Office, Command, Control, Communications, Computers, and Intelligence (PEO C4I)/Networks, Information Assurance and Enterprise Services Program Office (PMW 160), 2007.
RAND *MG907-2.1*

In the next two sections, we supply short descriptions of selected C2 and networks and communication systems used by the Navy. We do not include descriptions of Navy intelligence systems. Because this monograph is focused on potential changes in the design and construction of Navy ships to reduce the costs and increase the effectiveness of C4I upgrades, we are more interested in systems that process and exploit information than with those that generate information.

C2 Systems

C2 systems include software, methods, and procedures that enable commanders to make decisions and control their forces. C2 systems take information from intelligence systems and, with input from humans, aid in providing numerous C2 functions from the tactical to the strategic levels of war. Program Manager, Warfare (PMW) 150 is the C2 program office for C2 systems within PEO C4I.

Some C2 systems used by the Navy are widely employed within the U.S. armed forces more broadly. For instance, GCCS-M, the maritime version of the widely used GCCS, is installed and used on almost every ship in the Navy.[6] GCCS-M is a family of C2 systems that displays information on friendly, hostile, and neutral forces. This information is integrated with environmental and other information to support command decisionmaking. Future programs, such as Joint Command and Control, will follow GCCS-M as the next generation of C2 systems.

Numerous other C2 systems fielded within the Navy help commanders make decisions and control their forces. A study by the National Academies of Science provided an overview of programs available across the services that are used by the Navy (see Table 2.1). Navy C2 systems come from both service-specific applications and joint program offices.

Current trends in C2 systems within PEO C4I focus on applications that conform to open-architecture (OA) standards.[7] Applications for C2 functions developed quickly as the technology revolution advanced, and they can be complex and inflexible in terms of changes and upgrades. To counter this inflexibility, future naval C2 applications should conform to OA standards to help to simplify the relationships among C2 support systems and to decouple the data producers from the consumers. This latter objective will involve implementing appropriate mechanisms and rules to make certain that data are produced for any application rather than just one specific application. The decoupling will enable the cohesive development of particular C2 functions across the fleet and among other joint entities. One example of a case in which OA could make a positive impact by reducing data redundancy and particularity is in the generation of the common operational picture (COP). The COP is the result of the comprehensive collection and

---

[6]   See Department of Navy, Research, Development & Acquisition, undated, for additional program information. Note that GCCS-M itself has C2 decision aids and applications in the hundreds strung together into an integrated package.

[7]   Navy support for OA adoption was codified in Deputy Chief of Naval Operations, Warfare Requirements and Programs (N6/N7), 2005b.

Table 2.1
C2 Systems Used by the Navy

System Name	Abbreviation	Program Management Organization
Command and Control Processor/Common Data Link Management System	C2P/CDLMS	PEO C4I
Common Link Integration Processing	CLIP	PEO C4I
Expeditionary Decision Support System	EDSS	Marine Corps Systems Command
Global Command and Control System–Joint	GCCS-J	Defense Information Systems Agency
Global Command and Control System–Maritime	GCCS-M	PEO C4I
Joint Effects Model	JEM	Navy
Joint Interface Control Officer Support System	JSS	Air Force
Joint Operational Effects Federation	JOEF	Army
Joint Protection Enterprise Network	JPEN	Army
Joint Simulation System–Maritime	JSIMS-M	Navy
Joint Warning and Reporting Network	JWARN	Army
MIDS–Low Volume Terminal	MIDS-LVT	Navy
MIDS and F/A-18 Integration	MIDS F/A-18 Integration	PEO C4I
MIDS on Ship	MOS	PEO C4I
Naval Tactical Command Support System	NTCSS	PEO C4I
Theater Battle Management Core System	TBMCS	Air Force

SOURCE: Adapted from National Research Council, 2006.

fusion of ground, air, and maritime tracks; it allows all users, regardless of service or unit affiliation, to access similar underlying data to display their pictures. A COP will therefore need to be developed jointly and will be based on services specific to the demands of the users.

To develop a COP and other C2 applications with an OA, the Navy will also focus on service-oriented architectures (SOAs). The SOA concept involves providing loose coupling among software systems to allow for easy access to data, easy upgrades, and easy composition of the next product. The Department of Defense has subscribed to some of the tenets of SOA through the Defense Information Security Agency's Net-Centric Enterprise Services (NCES) acquisition program,[8] which will provide a number of common services (e.g., COP generation) across the forces. One result of the Net-Centric Enterprise Services program is the Joint Command and Control System, to which GCCS-M will transition in the coming years. The transition will allow users to more effectively employ information that comes from the joint community to address service-specific needs.

Networks and Communication Systems

Networks and communication systems support C2 functions and intelligence, surveillance, and reconnaissance through the processing and transport of information over physical media (such as NCES and fiber-optic cables) and through space via electromagnetic transportation. Within PEO C4I, PMW 170 is in charge of communications, and PMW 150 is in charge of C2 for the program office. PMW 160 is responsible for networks, including the Consolidated Afloat Networks and Enterprise Services (CANES) program.

The Navy communications infrastructure consists of satellite and terrestrial systems composed of copper, fiber-optic, and wireless communication media and associated receiver hardware and software to perform networking needs (see Figure 2.2).

The Navy's communications technical roadmap, which describes the future for communication programs, envisions reducing the

---

[8] For more information on NCES, see Defense Information Systems Agency, undated.

Figure 2.2
Communication Systems in the Navy

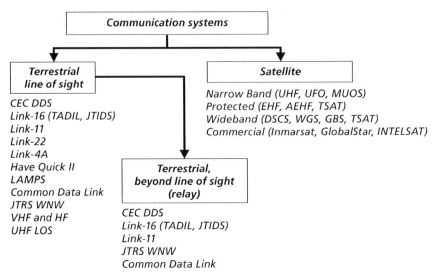

SOURCE: Adapted from National Research Council, 2006, p. 152.

RAND MG907-2.2

number of legacy communication systems and variants.[9] For instance, the future Navy Multiband Terminal will replace super-high frequency (SHF), extremely-high frequency, and Global Broadcast System terminals in 2012 and beyond.[10] PEO C4I estimates that nearly 900 variants of radios and communication equipment are fielded within the Navy, with some legacy systems (e.g., WSC-3) being over 30 years old.[11] The Department of Defense vision for communication systems is based on the Global Information Grid (GIG),[12] which is managed by the Assistant Secretary of Defense for Network Integration and Infrastructure.

---

[9] U.S. Navy Program Executive Officer, Command, Control, Communications, Computers, and Intelligence Team, 2007a.

[10] U.S. Navy Program Executive Officer, Command, Control, Communications, Computers, and Intelligence Team, 2007a, p. 7.

[11] Adan, 2007.

[12] See Department of Defense, Chief Information Officer, 2007, for further information on the GIG.

The Navy's FORCEnet concept is aligned with the GIG, and some programs currently under way within the Navy (such as the Navy/ Marine Corps Intranet) are being aligned with the GIG.

Naval forces have four primary shipboard infrastructure networks: the Non-Secure Internet Protocol Router Network, the SECRET Internet Protocol Router Network, the Sensitive Compartmented Information (SCI) Network, and the Combined Enterprise Regional Information Exchange System.[13] These networks constitute the backbone transmission hardware and software that allows information to flow into, within, and out of the Navy's platforms. Future networking infrastructure is expected to be consolidated further as a result of the large costs of maintaining interoperability and support for numerous different legacy systems developed over time. The different systems introduce a variety of nonstandard hardware and software requirements that pose security problems and raise costs.

PEO C4I has therefore provided a roadmap for consolidating network infrastructure into a minimum number of systems. This consolidation will

- reduce the number of installs required
- reduce the physical footprint of the set of systems
- allow computing power to be used more efficiently
- reduce the costs of managing configurations
- enhance security
- reduce nonrecurring engineering costs
- reduce manpower and training costs
- achieve economies of scale through commonality of racks and servers.[14]

The consolidation of communication systems entails migrating the numerous network infrastructures into a single overarching pro-

---

[13] One source notes that the average force-level ship might contain 50 separate networks. See Turner, 2007, p. 26.

[14] Adapted from U.S. Navy Program Executive Officer, Command, Control, Communications, Computers, and Intelligence Team, 2007a, p. 8.

gram called CANES. CANES will consist of the Integrated Shipboard Network System (ISNS) and of capabilities from other programs, such as the Combined Enterprise Regional Information Exchange Systems, the SCI LAN, and the Submarine LAN. ISNS, the Automated Digital Network System, GCCS-M, and the Naval Tactical Command Support System are current programs within PEO C4I.

The next chapter identifies the problems that arise when upgrading C4I systems and describes what the program offices and shipbuilders are doing to resolve those problems. First, however, we briefly discuss the difficulties associated with scheduling C4I upgrades.

C4I Upgrades and In-Service and New-Construction Timelines

Problems in fielding state-of-the-market C4I technologies in the Navy are well-documented.[15] Fielding must take into account the long period between deployments and depot maintenance periods. One such depot maintenance period is an aircraft carrier's midlife refueling complex overhaul (RCOH). Figure 2.3 shows the midlife RCOHs and in-service and retirement points for the current and projected aircraft-carrier fleet.

When aircraft-carrier RCOHs are staggered approximately four years apart, technologies deployed on one hull can become substantially different from those on the previous hull due to the rapid technology development typical of C4I systems. Similar timeline issues apply in new construction as well (see the bottom of Figure 2.3).

C4I system integration across the fleet over such long periods is made difficult by the rate of change of the individual technologies. The highly technological nature of the systems implies that refresh rates will be comparable to or faster than system build times. Use of COTS, government off-the-shelf, and nondevelopmental-items[16] hardware

---

[15] For additional information, see Iacovetta et al., 2000.

[16] Nondevelopmental items are previously developed items that only the government is allowed to purchase. Such items require only minor modifications to be usable to the government. COTS items, on the other hand, are previously developed items that the public is allowed to purchase.

Figure 2.3
In-Service and New-Construction Timelines

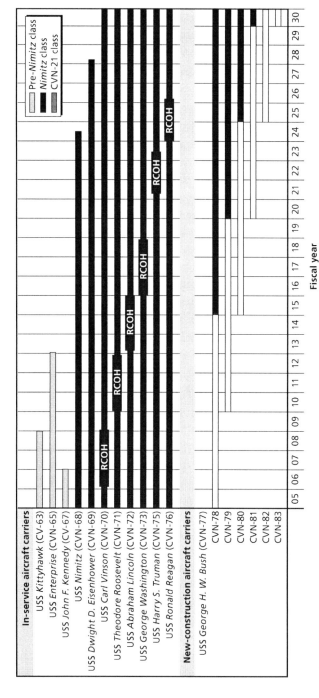

NOTE: All dates are approximate.

presents configuration-management challenges because some technology development is left to the commercial world and thus kept out of the government's control.[17] (There are benefits to this arrangement, however, including spiral introduction, commonality across platforms, and larger buys across systems.) Thus, for the Navy to take advantage of the full capabilities of the C4I systems it selects, acquisition and integration must specifically address the rate of change of the individual technologies in relation to the construction and rebuilding rates of the ships.

---

[17] Recent examples of COTS insertion into major Navy systems are discussed in Walsh, 2005, and National Research Council, 2004.

C4I Upgrades: Issues and Options

In this chapter, we describe the major issues involved in upgrading C4I systems and identify design and construction options, discussed during various interviews, aimed at overcoming those issues. These problems and options can be grouped into three categories: those that arise or apply during new ship construction, those that arise or apply for in-service ships, and those that are common to both new-construction and in-service ships.

C4I-Upgrade Issues During Ship Design and Construction

During the design and construction of new ships, there are four major issues related to the installation costs of C4I systems:

- adopting commercially available systems versus special-built systems
- deciding which C4I systems will be government-furnished equipment (GFE) and which will be contractor-furnished equipment (CFE)
- delivering ships to the fleet with the most-up-to-date systems
- incorporating adequate design margins for weight, power, cooling, and bandwidth into the ship design.

The first three issues are interrelated and primarily involve choices concerning the management of new-ship programs rather than the specific design and construction of ships. The fourth issue, design margins, is directly related to ship design but is also influenced by the fiscal aspects of a program.

Adapting Commercially Available Systems and Technologies to Shipboard Environments

The current naval fleet includes many ships and submarines designed 30 years ago. These vessels used to include C4I systems designed and built by a shipbuilder or systems organization using proprietary hardware and software. This resulted in the Navy becoming dependent on those shipbuilders or system developers for updating and modernizing the C4I applications. Locked within the boundaries and constraints of the chosen technology, the Navy made updates to hardware and software infrequently and often found them very costly.

Thirty years ago, the field of personal computing was just starting to grow, and the majority of people were still unfamiliar with computers and computer technology. But growth came rapidly as the cost of computing capabilities steadily declined and their use became more widespread in the marketplace. Hardware with capabilities unimagined or unaffordable 30 years ago is readily available at affordable prices today. Software with widespread utility is standard on today's personal computers (PCs). Networks tie together multiple computers and support equipment that is either collocated or decentralized geographically.

Navy ship programs, and the system designers and builders that support those programs, have taken advantage of the increasing capabilities and declining costs of commercially available computing hardware, software, and networking systems. Every newly constructed ship in today's fleet includes COTS systems, and ships originally fitted with special-built systems have been upgraded with COTS technologies. The same computers and networking capabilities used by commercial organizations, at all levels of the U.S. education system, and in the majority of American homes are also used by sailors on naval ships to accomplish everything from personal computing to controlling ship operations.

The Navy faces a few disadvantages, however, when it exploits the declining costs and increasing capabilities afforded by the commercial marketplace. COTS equipment is designed for commercial or home use, not for a shipboard operating environment. Peacetime naval operations subject the equipment to salt air and the pitching and rolling of the ship, and even-more-taxing demands occur in wartime environments, which require the equipment to operate successfully after sustaining significant shocks.

Military specifications for shock, quieting, and maritime operations can result in the addition of significant modification costs to the otherwise inexpensive commercial equipment. However, the submarine community has overcome these problems by isolating the equipment from the source of the potential shock. The "rafting" approach it uses for the design and build of the *Virginia*-class submarine basically "floats" the equipment above the outside structure of the submarine. This not only allows for the use of commercial equipment with minimal modification but also permits the equipment to be easily upgraded without significant installation costs. The DDG-1000[1] program features a similar strategy to isolate its C4I equipment for shock and quieting reasons.

The downsides of modular isolation are (1) the requirement for special mounts for the isolated deck structure and (2) the additional weight the approach can add to the overall deck and ship. Also, rafting requires additional development and testing to demonstrate shock and vibration sufficiency. The *Virginia* program overcomes some of these disadvantages by testing the integrated modules in a specialized facility before they are placed into the submarine during construction.

A second disadvantage to the Navy of using COTS equipment in the design and construction of its ships is the rapid development and refresh cycle found in the commercial marketplace. Software "turns over" every two to three years, and hardware experiences major upgrades or improvements every five to six years. Older hardware and software are still very usable and can typically provide the desired capabilities, but obtaining support for either repairing or replacing components in the hardware or overcoming a software problem becomes increasing difficult and costly as these elements age. The rapid refresh rate of commercial technology, coupled with the long time required to design and build new ships and to provide upgrade opportunities to in-service ships, hampers the Navy's ability to have standardized C4I sys-

---

[1]   A DDG is a guided missile destroyer.

tems across the fleet and even across a specific class of ships. As a result, program managers (PMs) face tough decisions about what specific C4I COTS technologies to include during ship design and construction, an issue discussed in more detail in the next section.

Although COTS technologies are widely used in ship-design and ship-construction programs, they lead to another issue: Who should be responsible for providing the C4I systems, the Navy or the ship designer or builder?

Deciding on GFE Versus CFE Systems

Currently, there is no preferred provider of C4I systems for new ship programs—each new program can choose to either use the systems provided and supported by PEO C4I or ask the shipbuilders competing for the ship design and construction to develop their own C4I solutions. Often, the end result is a mix of GFE and CFE for the suite of C4I equipment and software on a ship. Systems that are unique to a class of ships are typically CFE, while those that are common across multiple ship classes are typically GFE.

There are advantages and disadvantages to both GFE and CFE systems. GFE equipment allows standardization across the fleet, and such equipment is easier to support once a ship enters the fleet. Complete standardization across the fleet of even one C4I system is probably an unattainable goal, however, because of the high refresh rates of commercial technologies and the varying timelines that make ships available for upgrades. Moreover, ship PMs argue that the PEO C4I systems are often a generation behind what is available in the open market and that ship designers or the companies that design and build the C4I hardware, software, and systems can provide equal or better capability at a lower cost.

CFE systems can leverage the broader commercial marketplace and provide multiple technology options, hopefully at a lower cost. However, CFE systems can present a problem to PEO C4I once a ship enters the fleet and responsibility for the support of the systems transitions to the Navy, which may not have complete knowledge of the systems. Also, CFE systems may create a unique logistics tail, requiring a separate spare-parts pool and system-unique training for the sailors who operate and maintain the system. This was a major problem with the Shipboard Wide Area Network, which was developed by a ship-builder for the LPD-17 class of ships.[2] Once those ships entered the fleet, PEO C4I had difficulty understanding the nuances of the system and, therefore, had difficulty upgrading and supporting the system. Current plans call for replacing the Shipboard Wide Area Network with CANES on the ships in the LPD-17 class.

Whether GFE or CFE for C4I systems is used varies by program. The DDG-51 program uses systems provided by PEO C4I, while the *Virginia* program uses several CFE systems. The issue of whether PEO C4I should be the preferred provider for new ship programs warrants further study, especially in terms of the total life-cycle cost impact of choosing either GFE or CFE.

Ensuring the Currency of C4I Systems at Delivery

One of the biggest problems facing ship PMs during new-ship construction is ensuring that the most-up-to-date C4I technologies are incorporated in a new ship at delivery. C4I technologies can refresh multiple times during the several years it takes to build a ship. If C4I systems are identified too early in the ship-construction process, the systems may need to be removed and replaced as soon as the ship is delivered to the Navy. On the other hand, ship designers and builders need to lock in both the dimensions and foundations required for the C4I systems and the systems' power and cooling requirements as early as possible in the design and construction process. If specific equipment is not identified early enough, sufficient space, power, and cooling may not be available to support the equipment.

It is difficult to avoid both locking in systems too early and waiting too long to identify specific equipment. Certain C4I systems are needed early during ship construction to support the onboard crew and to adequately develop the overall test and integration programs. Total integration and testing can only occur when all the systems are available on the ship. Also, if specific systems are not identified early enough, the equipment delivered for integration into the ship may not

---

[2] An LPD is an amphibious transport dock.

match the engineering detail specified during ship design. Finally, the Navy must meet the regulatory requirement of delivering a fully operational ship, and it therefore requires all systems to be installed and working correctly when the ship is delivered.

Ship PMs have developed a strategy to overcome the problem of obsolescence in C4I equipment at ship delivery. The strategy—termed *design budget, technology insertion,* or *turnkey,* depending on the program—basically provides ship designers and builders the broad specifications of the C4I equipment in terms of space, power, and cooling requirements without identifying the specific equipment that will be installed. Technical data are provided to shipbuilders at specified times during the build process to allow the C4I spaces to take shape. Not-to-exceed values for space, power, and cooling are defined, and ship managers must stay within these parameters to ensure that there are no problems installing the equipment when it is finally specified. Often—in the case of submarines and surface combatants, for example—C4I equipment is installed on racks and tested before the racks are integrated into the C4I spaces on the ship.

The *Virginia* class of nuclear attack submarines is an example of a current build program during which attempts to ensure the currency of C4I systems and the provision appropriate spaces at delivery have proven successful. The Navy defined the space, power, and cooling requirements for the shipbuilder, but the shipbuilder has maintained ownership of the racks and the specific equipment installed on the racks. The shipbuilder uses installation teams to install and test the equipment shortly before the submarines are delivered to the Navy.

Normally, the Navy will, after a ship is delivered, conduct demonstration and shakedown cruises to test the ship's systems and identify any manufacturing shortcomings. These issues are then addressed in a postshakedown availability (PSA), which often includes a refresh of C4I systems. Subsequently, crew training is conducted, and ships are typically available for operations approximately two to three years after delivery. The C4I-management program used for the first ship of the *Virginia* class allowed this practice to be reversed. The USS *Virginia* was the first submarine to be deployed before its PSA. Therefore, the submarine underwent its first technology refresh just before delivery and its second during the PSA, which occurred more than a year later.

Other techniques are being used to reduce the amount of "hot work" (as welding and the installation of foundations and structures are known) that is needed to upgrade C4I equipment both during new construction and over the service life of the ship. Standardized racks for holding C4I equipment are replacing specially built enclosures, allowing C4I boxes to be replaced without requiring the removal of old enclosures and foundations and the installation of new ones. The CVN-21 aircraft-carrier program[3] is including a flexible infrastructure for C4I spaces whose modular power and cooling connections and walls mounted on tracks will allow for the quick reconfiguration of C4I spaces. These two types of options to reduce the cost of hot work when updating C4I system are discussed later in this chapter.

Incorporating Adequate Design Margins

The solution for facilitating C4I upgrades that was most often mentioned during interviews was setting and maintaining adequate design margins. The design process for all naval ships includes extra ship capacity for weight, power, and cooling. Adequate design margins must also reflect considerations for additional bandwidth on the ship. These design margins are intended to allow the ship to incorporate additional capabilities over its service life. Initial design margins are typically on the order of 10–20 percent of the total, with submarines typically featuring the larger design margins. However, extra capacity leads to increased ship-construction costs, and design margins are often reduced when budgets become constrained. Also, design margins are often set very early in the design process, and new C4I technologies often become available during the design and build processes. These additional systems typically eat into the design margins before the first ship in the class is ever delivered to the fleet.

The majority of the ships in the Navy fleet have little extra capacity to meet increased demand for weight, power, cooling, or bandwidth because the initial design margins are typically consumed very early

---

[3] A CVN is an aircraft carrier, nuclear.

in a ship's life. This lack of extra capacity results in difficulties in finding ship services to support the new equipment and in additional costs when upgrading C4I capabilities, especially when adding new capabilities to a ship. Workarounds, including the addition of new power and cooling plants, are often needed to provide the additional ship services required to accommodate C4I upgrades. Existing systems may have to be downgraded, or ship operations may need to be constrained (e.g., by not operating certain systems while others are being operated). Some ship classes have become so constrained that a new system or capability cannot be added without an existing system being removed. The submarine community seems to do a better job of maintaining margins than do the communities of other classes of ships.

The issue of adequate design margins is being addressed in the design of new classes of ships. For example, the design for the new *Ford* class of aircraft carriers includes both extra cabling in various spaces and extra bandwidth. The design also features a zonal electricity grid that allows power to be directed throughout the ship where and when it is needed.

C4I-Upgrade Issues for In-Service Ships

Many of the issues that arise during ship design and construction also arise when upgrading the C4I systems of ships that are in the operational fleet. There are, however, two problems that are unique to in-service ships: (1) the variable configurations of individual ships within a class of ships and (2) getting approval for C4I upgrades through the ship maintenance (SHIPMAIN) process.

Various Ship Configurations Make the Planning of Upgrades Difficult

The cost of specific upgrades to C4I systems could be reduced if the installation details were the same across all ships in a given class. This uniformity would permit both the creation of just one set of design drawings for the upgrade and the implementation of a repetitive approach to installing the upgrades. Unfortunately, the C4I systems of various ships in a given class feature different configurations, and the areas where C4I systems are installed may be differently laid out. This problem is especially acute in those classes that have long production runs (e.g., the DDG-51 class) or when several years intervene between the construction of each new ship in the class (as in the case of *Nimitz*-class aircraft carriers). Nonstandard configurations across the ships in a single class mean that each installation requires a check of the ship's configuration, the development of a unique set of design drawings, and an almost-unique process for installing the upgrade. These requirements contribute to increased cost and time to accomplish the upgrade.

Several factors contribute to various ships in a class having different C4I configurations. In some cases, the production of the ships in the class can span two decades or more, and design changes can occur over the length of the production run. For example, the DDG-51 class has featured various ship configurations over its more than 20 years in production, and the *Nimitz* class has also seen multiple changes over the course of its 40 years in production. In some cases, ships in a single class are built at more than one shipyard, which can result in different ships exhibiting slightly different configurations of the C4I systems or the layouts of the areas devoted to those systems. However, the biggest contributor to different configurations is the high refresh rate of the technology used in the C4I systems. For example, the C4I configurations in DDG-51–class ships currently in construction will differ from those in DDG-51–class ships delivered only a few years ago. Because operational demands often limit when technology upgrades can be installed on a ship, it is virtually impossible to maintain a constant configuration of C4I systems within a class.

The submarine community appears to do the best job of maintaining similar C4I configurations across all of the submarines within a class and even across various classes. Part of this success is due to adapting an OA design philosophy and business model for C4I and combat systems. The Acoustic Rapid COTS Insertion program has established a process of identifying and prioritizing fleet requirements and developing the C4I changes needed to meet those requirements. New hardware baselines are developed every two years, and the objective is to replace hardware on each attack submarine every four years. Software is upgraded every two years.

The submarine community strives to maintain the same configuration on every ship. One example is the goal to maintain a common radio room on each submarine. The objective is to have a common layout and equipment configuration for the radio room on all submarines so that any sailor can go into the radio room on any submarine and immediately be familiar with the equipment and operations. This practice greatly reduces C4I-system training and documentation costs and is so successful that there are now plans to maintain a common radio room on surface ships.

Navigating the SHIPMAIN Process

The SHIPMAIN process was initiated in November 2002 to transform the maintenance and modernization process for Navy surface ships.[4] It seeks to identify and eliminate redundancies in maintenance processes so that the right maintenance is done at the right time, at the right place, and at the right cost. It also seeks, through a common planning process for ship maintenance and modernization, to maintain configuration control of the various changes made to ship systems and equipment over the life of a ship.

For modernization activities, such as C4I upgrades, SHIPMAIN instituted a single, disciplined process that allows various alterations (i.e., modernizations) to be ranked and prioritized through an approval process involving multiple stakeholders. The process sets three decision points: one related to funding preliminary engineering studies, one related to funding the design and development of the alteration plans, and one related to funding the accomplishment of the alteration. The objective is to procure and install approved alterations within three years.

SHIPMAIN has been successful in reducing the churn in maintenance and modernization planning and in reducing maintenance costs while providing ready ships to the fleet. It has also been successful in establishing and sustaining configuration control across the

---

[4] Submarines had their own process, Fleet Modernization, in place prior to the introduction of SHIPMAIN.

ships in a given class for hull, mechanical, and electrical equipment and systems. However, because C4I technologies can change one or two times during the three-year planning-and-implementation period, SHIPMAIN is typically viewed as an impediment by the organizations responsible for keeping the most-up-to-date computing and communications technologies on in-service ships.

These organizations feel that the SHIPMAIN process is too difficult and too-time consuming to implement during C4I upgrades. The process—and the time required to approve and implement changes— is the same regardless of the magnitude of the change.[5] For example, software upgrades go through the same SHIPMAIN process as major hardware changes. Software upgrades are also more difficult than hardware changes to certify in the SHIPMAIN process because it is difficult to specify the impact of a software change from an operational perspective. There is typically a significant integration-and-testing phase associated with C4I upgrades, and if the upgrade does not function as planned after the testing, the SHIPMAIN process must be implemented yet again to make changes to the original upgrade.

In interviews, various organizations noted that, with waivers, an alteration can be approved within 90 days under SHIPMAIN. However, the approval process itself takes time and resources. Because quick approvals are typically required for C4I upgrades, this shorter alternative should be streamlined and made easier to navigate.

C4I System Upgrade Issues Common to New-Ship Construction and In-Service Ships

A few factors influence C4I installation costs during both new-ship construction and when upgrading in-service ships: the need to integrate and test the new systems installed in the ship, the amount of hot work and changes to ship services required, and the need to integrate the antennas on the topside of the ship.

---

[5] Process improvements have helped correct this problem in some areas, such as engineering changes and fielding changes.

Integration and Testing

The proliferation of C4I systems on naval ships has complicated the process of integrating the various systems into an overall ship C4I architecture and testing the overall C4I package to ensure all functions are working correctly. New C4I capabilities beyond those originally planned during the design stage were added in a piecemeal fashion to many classes of ships currently in the fleet. These new capabilities not only strain the ability of the ship to provide power, cooling, bandwidth, and other support functions but have also proved complex to integrate with existing capabilities on the ship. Often, four or five different networks on the ship would compete for support from ship services and require a complex integration-and-testing plan.

How systems integration will occur is a key decision in the C4I design process. The federated approach, used successfully by the submarine community for the *Virginia*-class program, decentralizes the hardware and software functions of the various C4I systems while allowing all systems to share data and information through a common network. Under this decentralized approach, the hardware or software problems of one C4I system, or of an upgrade to a system, can be isolated and addressed without affecting the performance of other C4I systems. Another advantage of the federated approach is that it widens the available pool of potential vendors of the C4I capabilities. Various C4I systems can be procured from different providers and "plugged into" the federated system through a common set of specifications.

One downside of using a federated approach is that doing so involves some level of redundancy in C4I hardware and software, which results in increased costs and increased demands on ship services. Another disadvantage is the need to establish and maintain a set of rigid specifications that governs the connection of the C4I systems into the federated architecture.

An integrated architecture reduces hardware and software redundancies by using both a shared network and shared computing hardware and system software. This approach was used in early *Ohio*-class submarines and is being used in the current CANES approach. C4I functions in integrated systems are typically supplied by software programs that use common system-hardware processing and the common network to share data and information. The disadvantages of integrated systems are (1) the potential loss of all C4I functions when a problem with the common hardware or system software arises and (2) the fact that there are few suppliers capable of delivering the more-complex hardware and software systems involved in this type of architecture.

Regardless of the overall architecture chosen for C4I systems, a consolidated testing plan is needed to ensure that all C4I systems are working correctly in both a stand-alone mode and as part of the overall C4I architecture. A consolidated testing plan is developed during the construction of the ship, but it should also serve as a guide for updating C4I capabilities throughout the life of the ship. The plan is an agreement between the Navy, the shipbuilder, and the C4I suppliers that specifies when and how testing will be accomplished. The overall objective of the plan is to allow equipment to be tested once in a manner that is accepted by all parties.

A consolidated testing plan should reduce redundant activities, thus saving money, and allow testing to occur later in the build process. However, a consolidated testing plan typically requires that a facility perform the testing of the overall C4I system, and constructing and operating such a facility entail monetary, schedule, and opportunity costs. Also, a consolidated plan requires the cooperation and coordination of the various parties involved.

The *Virginia* program has successfully implemented a consolidated testing plan in which the electronic components of the C4I weapon systems are assembled and tested at the Command and Control System Module Off-Hull Assembly and Test Site facility at Electric Boat before they are inserted into the submarine. The Aegis combat system undergoes a similar process before its insertion into a DDG-51–class hull.

Hot Work and Changes to Ship Services

The need for hot work and changes to ship services (i.e., space, power, and cooling) drives the costs of C4I upgrades. If adding a new capability to or upgrading an existing capability on a ship were as easily accomplished as upgrades to home computers or audio-visual systems, C4I-upgrade costs would not be an issue for the Navy. However, the complexity of integrating C4I systems, the limited supply of ship services, and the density of modern ships require a significant amount of labor to remove and replace equipment.

The Navy is employing several initiatives to reduce the required amount of hot work and changes to ship services. For example, the CVN-21 aircraft-carrier program is incorporating a flexible infrastructure into various C4I spaces on the *Ford* class, and changes since the last design include a raised deck with ventilation running underneath and movable vents on the deck to direct the cooling where it is needed. The walls are mounted on tracks to allow for quick reconfiguration of the spaces. The deck also incorporates movable rails for securing the racks and equipment. Flexible power connections allow for installation and movement of equipment without rewiring the spaces.

A type of flexible infrastructure called *Smart Track* was originally included on the LPD-17 class. However, this implementation added additional weight and cost to the construction of earlier ships. Therefore, future ships in the class will not use the Smart Track system. The CVN-21's flexible infrastructure uses lightweight aluminum (rather than steel) for the decks to help control the additional weight. A business case that assessed the total life-cycle–cost implications of using the flexible infrastructure substantiated the potential cost savings, which are due to less labor being required for hot work and for running cables and ducts during C4I upgrades. Flexible infrastructure is now being considered for other new classes of ships.

New-ship designs are also incorporating features to allow for easier access to C4I systems during upgrade installations. Wider passageways and less-dense placement of C4I equipment allow for the removal and installation of new equipment without affecting other equipment. Wider access paths and technology-insertion corridors on the *Ohio* class allow equipment to be quickly repaired and systems to be quickly upgraded so that the submarines can adhere to a quick operational-turnaround cycle.

New-ship designs are also using standard racks for C4I systems. These standard racks allow for easier removal and replacement of equipment without the need for extensive hot work to remove old foundations and install new ones. As a result, a ship's crew can often accomplish C4I upgrades while the ship is at the pier rather than having to send the ship to an industrial facility for the work to be done by an installation team. The racks are based on commercial standards and facilitate the use of COTS equipment.

A potential downside to the use of standardized racks is the difficulty of defining and maintaining the "standard." COTS equipment has become smaller as new microprocessor technologies are developed, and this has made a standard size difficult to define.

Topside Integration

Integrating antennas for the various C4I systems on the topside of the ship is often more difficult than integrating C4I systems within the ship structure. Each antenna must have a clear field of view to receive signals, and it cannot cause electronic interference with other antennas. Topside space, from both the horizontal and vertical perspectives, is limited and must be carefully managed. There are efforts under way to consolidate various antennas, but these efforts have not yet led to an acceptable solution. The submarine community has made progress with the topside integration problem by adopting a universal modular mast that allows antennas to be changed more easily.

Summary

This chapter has described the various C4I-upgrade problems that arise during the design and construction of ships and after ships have entered the fleet. It has also reviewed the program-management decisions and design alternatives adopted to mitigate these problems. Table 3.1's columns summarize these various challenges, and its rows list the design and management options aimed at lessening the impact of these problems during C4I upgrades. As the table shows, single management decisions and design options are typically aimed at solving multiple upgrade problems.

Some of the actions are aimed at solving C4I-upgrade problems during new construction, others address problems that arise during upgrades for in-service ships, and some are appropriate in both cases. Table 3.2 shows these relationships.

Table 3.1
Actions Taken to Solve Various C4I-Upgrade Problems

Action Taken	Cost of Hot Work	Ship Service Modifications	Military Specifications	Systems Integration or Testing	High Tech Refresh Rate	Requirements Growth
Create a flexible infrastructure	X	X			X	X
Employ standard racks or enclosures	X	X			X	X
Employ modular isolation			X			
Improve access to C4I-system spaces	X	X			X	
Use the design-budget process during construction					X	X
Incorporate adequate design margins					X	X
Decide on federated vs. integrated systems			X	X		
Employ consolidated testing				X	X	
Use COTS, SOA, and/or OA					X	X

Table 3.2
Actions Aimed at Different Phases of the Ship Life Cycle

Action	Design or Build Phase	In-Service Ships
Create a flexible infrastructure		X
Employ standard racks or enclosures		X
Employ modular isolation	X	X
Improve access to C4I-system spaces	X	X
Use the design-budget process during construction	X	
Incorporate adequate design margins		X
Decide on federated vs. integrated systems	X	X
Employ consolidated testing	X	
Use COTS, SOA, and/or OA	X	X

CHAPTER FOUR

Analysis of C4I Installation Costs

To help lower the installation costs of C4I components, one must be able to predict and track costs accurately and precisely understand what factors drive both installation costs and variability in those costs across and within different classes of ships. Once the significant cost drivers are found, steps can be taken to reduce their impact on actual cost, or their effects can be included in future estimates. In this chapter, we focus on questions related to C4I-upgrade cost drivers, cost variability, and estimate accuracy.

Cost estimating[1] is a core competency of any program organization and an essential part of the budgeting-and-planning process. Accurate estimates allow an organization to make appropriate trade-offs between competing priorities. Errors in cost estimates can be detrimental to efficiency: Overestimates tie up money that could be used for other projects, and underestimates can result in budget shortfalls. Cost estimating serves a secondary purpose: project control. Estimates serve as the comparison baseline. Without an accurate baseline, monitoring the execution of a program becomes difficult. With good estimates, an organization can monitor the progress of programs and determine whether there are problems that might need to be mitigated.

We conducted a detailed analysis of the financial data on the installation of a number of C4I upgrades that took place sometime in 2000–2008. This chapter begins with a description of our general methodology for analyzing the data. Next, we describe four specific

---

[1] By the term *estimating* we mean the entire cost process—from the generation of estimates through the tracking and collection of actual costs.

systems (some of which include more than one type of upgrade) that we were asked to examine in terms of the variability in C4I installation costs. In the next sections, we explore potential drivers of installation-labor cost and estimating accuracy. In the final section, we summarize our observations.

Methodology

PEO C4I asked RAND to explore various issues pertaining to C4I installation cost, including

- the variability of in-service upgrade costs by platform type
- whether where the installation is done—i.e., in a private shipyard, a public shipyard, or at the pier[2]—has an effect on cost
- whether cost improvement is evident in a given type of installation (i.e., where there is a learning effect)
- whether installation costs are different in the case of early adopters of CANES.

PEO C4I provided RAND a data set of nearly 12,000 system installations that took place sometime in 2000–2008.[3] Some of the important fields included in the data set were hull type, ship identification (i.e., hull name), coast where the work was done, upgrade type (i.e., a brief description of the upgrade), and availability type.[4] In addition to information on the actual installation-labor cost and start date, the data set included several estimates of the installation-labor cost from

---

[2]   We were unable to directly examine whether the location of an installation affects cost; instead, we examined availability type and determined on which coast (Atlantic or Pacific) the installation was accomplished.

[3]   The data were from the Space and Naval Warfare Systems Center/PEO Integrated Data Environment and Repository database.

[4]   By *availability type* we mean the status of the ship when the installation occurred. The two types are (1) a CNO availability (e.g., a planned incremental availability, a selected restricted availability, a docking planned incremental availability, or a docking selected restricted availability) and (2) a window of opportunity (WOO).

different sources. In our discussion of estimate accuracy, we compare the estimate generated by PMW, the PEO C4I organization responsible for producing the estimate (usually the first to be produced), to the actual installation-labor cost. When the actual cost of an installation was not available, we used the estimate at completion. When neither the actual cost nor the estimate at completion was available, we omitted the data point from the analysis. Some of the actual cost numbers seemed anomalously low (i.e., the entire installation was a few hundred dollars or less). Whenever these anomalous data affected our results, we inserted a comment to that effect. If the actual start date was not available, we used the latest of the estimated start dates; in other words, in our analysis, the updated estimate value took precedence over the original estimate value. If no information on dates was provided, or if the date information appeared to be erroneous (e.g., the start date was later than the date on which the data set had been compiled), the data point was omitted from the analysis, unless otherwise stated. Table 4.1 shows the number of C4I installations by year in 2000–2008 and dis-

Table 4.1
Number of Installations and Their Average Cost, by Year

Year	Number of Installations	Total Cost (FY09$ millions)	Average Cost (FY09$ thousands)
2000	17	1.7	100.3
2001	685	171.7	250.6
2002	549	124.6	227.0
2003	748	137.9	184.3
2004	1,333	168.3	126.2
2005	1,478	147.6	99.8
2006	2,131	228.8	107.4
2007	1,231	162.2	131.6
2008	296	52.1	175.9
Total	8,469	1,194.7	141.1

plays their average cost. Table 4.2 shows the installation data by hull type over the same period.

One question of interest is whether installation costs vary by hull type. Because hull type, one of the cost drivers we assessed, is a qualitative measure, we used the weight and length of the platform as a surrogate for that measure.[5] Table 4.3 summarizes the median hull length and median hull weight by hull type. (Note that we used the actual values, not the median values, for each hull in our analysis.) We found that hull length and weight are highly correlated; therefore, in the remainder of this monograph, we use hull length when looking at comparisons *between* hull types. (Hull length varies less within a given hull type than weight.) Weight is used as a comparison measure *within* hull types because it does vary within a single hull type.

Table 4.2
Number of Installations and Their Average Cost, by Hull Type

Hull Type	Number of Installations	Total Cost (FY09$ millions)	Average Cost (FY09$ thousands)
CG	1,113	139.7	125.6
CVN	762	135.8	178.3
DDG	2,088	225.0	107.8
FFG	792	59.2	74.7
LCC	153	28.5	186.2
L-class	1,767	255.1	144.4
SSN/SSBN	1,794	351.3	195.8
Total	8,469	1,194.7	141.1

---

[5] The original data set did not include information on ship weight, hull length, and commission date. We found these data in NAVSEA Shipbuilding Support Office, date not available.

Table 4.3
Median Hull Length and Median Hull Weight, by Ship Type

Hull Type (Abbreviation)	Length (ft)	Light Ship Displacement (tons)
Attack submarine, nuclear (SSN)	362	5,789
Guided missile frigate (FFG)	453	3,714
Guided missile destroyer (DDG)	505	6,740
Guided missile cruiser (CG)	567	7,096
Amphibious transport dock (LPD)	570	9,589
Landing ship dock (LSD)	610	11,332
Amphibious command ship (LCC)	635	13,038
Amphibious assault ship (LHA)	820	25,982
Amphibious assault ship (dock) (LHD)	844	28,050
Aircraft carrier, nuclear (CVN)	1,092	78,453

Cost Normalization

All actual and estimated installation costs were adjusted to constant fiscal year (FY) 2009 dollars using deflators from the Office of the Under Secretary of Defense.[6] We used the civilian pay index because this index is the closest analogy to installation labor. Because we knew the spending start and end points, we used the outlay index found in Table 5.9 of the Green Book.

Analytic Approach

We used techniques from statistics and data mining to look for significant patterns and other correlations in the data set.[7] We then tested the significance of potentially significant relationships using statistical analyses that primarily involved correlations, regression, t-tests, and

---

[6] Office of the Under Secretary of Defense (Comptroller), 2008. This reference is known as the *Green Book*.

[7] Most of the statistical work was done in MS Excel or STATA, and the data mining was done using the Waikato Environment for Knowledge Analysis software package.

analysis of variance (ANOVA).[8] Whenever a difference was found to be statistically significant, we reported the power of the significance.[9]

C4I System Upgrades Display a High Degree of Installation-Cost Variability

We were asked to (1) examine the variability of the installation costs of and (2) determine whether cost improvement was evident in upgrades to four specific types of C4I systems: Ships Signal Exploitation Equipment (SSEE), SHF radios, the ISNS, and the GCCS-M. Specific upgrades to these four systems include the following:[10]

- The SSEE Increment E upgrade provides surface platforms a comprehensive capability for tactical information, warfare exploitation, and electronic-warfare support measures.
- SHF Enhanced Bandwidth-Efficient Modems (EBEMs) for WSC-6 Variants provide surface ships with North Atlantic Treaty Organization interoperability and interoperable worldwide communications for naval and joint warfighting.
- Upgrades to ISNS supply the shipboard network infrastructure for all C4I and business applications. Three upgrades to the ISNS were analyzed: Embarkable Drops, LAN GIG–Electronic (GIG-E), and Common PC Operating System Environment (COMPOSE).

---

[8] Correlations can be used to determine how closely related two variables are. Regression can be used to find relationships between two or more variables. Both t-tests and ANOVA are used to determine whether the differences between subgroups of data are significant; T-tests are used for two subgroups, and ANOVA is used for more than two subgroups.

[9] The significance level is reported with a p value. For example, p = 0.95 indicates that the data suggest that there is a difference in means with a confidence of 95 percent (i.e., there is less than a 5-percent change that the difference detected is erroneous). Additional information on these approaches can be found in any statistics or econometrics textbook, such as Greene, 2003.

[10] A description of each of the upgrades to each of the four systems is provided in Appendix B.

- GCCS-M 4.X upgrades the Navy's fielded C2 system. This system provides the warfighter a current status report of vital positional information and other data needed to make tactical decisions.

Table 4.4 displays basic labor-cost statistics for installing these six specific upgrades. The second column in the table shows the number of installations we examined for each upgrade type. The remaining columns provide the minimum, maximum, and average costs along with the values of the lower quartile (the value at which 25 percent of the observations are less expensive) and the upper quartile (the value at which 75 percent of the observations are less expensive). These last two values show the range of costs for the middle 50 percent of the installations.

Table 4.4
Basic Labor-Cost Statistics for Analyzed Upgrades

		Labor Cost (FY09$)				
Upgrade Type	Number of Installations Examined	Minimum	25th Percentile	Average	75th Percentile	Maximum
GCCS-M GENSER 4.X (V) 1–4	31	55,469	158,121	283,651	459,345	843,693
ISNS Embarkable Drops	34	203	55,016	186,600	188,751	1,209,196
ISNS LAN GIG-E	54	33,356	1,029,918	1,577,444	1,624,571	8,268,285
ISNS COMPOSE 3.0 Software	82	249	27,947	43,303	51,617	165,532
EBEMs for WSC-6 Variants	48	13,550	32,212	42,929	50,682	82,439
SSEE Increment E	24	425,749	545,300	734,060	918,195	1,418,381

The values in Table 4.4 show the high cost variability of the different upgrades. We now turn to identifying potential causes of this variability.

Factors That Influence Installation-Labor Costs

First, we first examine the average installation-labor costs across platform types; then, we look at each platform to determine whether cost variability was a factor of either the age of the ship or the coast where the upgrade was installed.

Influence of Ship Type on Installation-Labor Cost

Table 4.5 summarizes the analysis of cost variability on different types of platforms. In general, there was a direct correlation between the labor cost to install the upgrade and the size of the ship (i.e., upgrades on larger ships cost more) for the following upgrades:[11] GCCS-M, ISNS Embarkable Drops, ISNS LAN GIG-E, and SSEE Increment E. However, there was an inverse correlation (i.e., upgrades on larger ships cost less) for the EBEM upgrade, and there was no apparent correlation between installation-labor costs and ship type for the ISNS COMPOSE upgrade. However, the coefficient in the linear model was positive, which indicates that, although it was not the primary driver, ship size did influence the installation-labor cost of the ISNS COMPOSE upgrade. When we found a positive relationship between hull size and installation-labor cost, we noticed that the CVN class of ships was generally involved. (CVNs are the largest class of ships in the data set, and they also tended to have the highest installation-labor costs.) In

---

[11] There are many reasons why installation-labor cost might correlate with size. With some installations, a larger ship might involve more terminals, longer cable runs, more servers, etc. These increased quantities would then translate into increased labor required to perform the installations. Conversely, installation could be more expensive on a smaller ship if access were more limited or shared infrastructure (e.g., chilled water) had less excess capacity. Although we note correlations with cost, we are unable to tie these trends to installation characteristics due to lack of data.

Table 4.5
Relationship Between Installation-Labor Cost and Ship Type

Upgrade Type	Number of Installations Examined	Correlation with Hull Length	Significance of Coefficient
GCCS-M GENSER 4.X (V) 1–4	31	Positive	p = 0.82
ISNS Embarkable Drops	34	Positive	p = 0.94
ISNS LAN GIG-E	54	Positive	p = 0.83
ISNS COMPOSE 3.0 Software	82	Positive (weak)	p = 0.99
EBEMs for WSC-6 Variants	48	Negative	p = 0.99
SSEE Increment E	24	Positive	p = 0.99

the next six sections, we discuss these findings in more detail for each of the six upgrades.

GCCS-M GENSER 4.X (V) 1–4. The average labor cost to install the GCCS-M upgrade on larger ships (i.e., the LHDs and CVNs) was significantly greater than the average labor cost to install the upgrade on smaller naval combatants (i.e., the DDGs and CGs). Figure 4.1 shows the average installation-labor cost by platform type. The average labor cost for installation on the DDGs was not significantly different from that of the CGs. The sample sizes for the LHDs and CVNs were too small to allow us to draw statistical conclusions about differences in their average cost.

The specific work package for the GCCS-M GENSER 4.X 1–4 upgrade was different for each class of ship (see Appendix B). Different systems were replaced, and different numbers of new components were installed. Therefore, differences in installation-labor costs among the

Figure 4.1
Average Labor Cost to Install the GCCS-M GENSER 4.X (V) 1–4 Upgrade,
by Hull Type

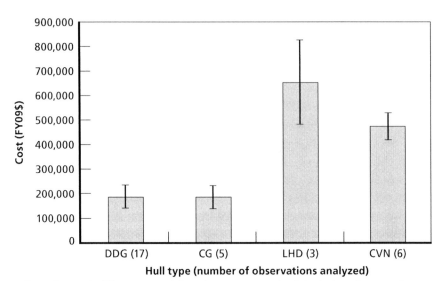

NOTE: The vertical line over each bar shows one standard deviation above and
below the mean.
RAND *MG907-4.1*

hull types may be due both to the presence of legacy systems on the
various hull types and to the size of the ship.

ISNS Embarkable Drops. The average labor cost to install the ISNS
Embarkable Drops upgrade appeared to increase on larger ships. This
trend is shown in Figure 4.2. Statistically, there was a strong correlation (0.76) between cost and hull length. However, the large variation
in installation-labor cost within the various hull types and the small
sample size available for certain hulls made it difficult to draw any statistical conclusions about differences in the average installation-labor
cost across hull types.

ISNS LAN GIG-E. The average labor cost to install the ISNS LAN
GIG-E upgrade on naval combatants (i.e., FFGs, DDGs, and CGs)
appeared to increase as the size of the ship increased. However, the
average installation-labor cost for LSDs was lower than that for the surface combatants. Also, the labor cost to install the upgrade on CVNs

Figure 4.2
Average Labor Cost to Install the ISNS Embarkable Drops Upgrade,
by Hull Type

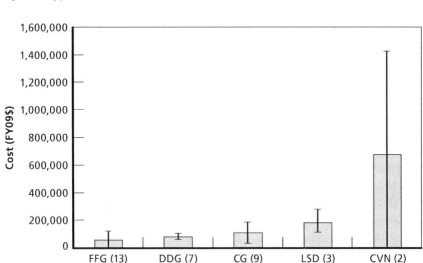

NOTE: The vertical line over each bar shows one standard deviation above and
below the mean.
RAND MG907-4.2

was greater than the cost for doing so on the other ship types, although
the small sample size prohibited us from drawing any statistical conclusions. Figure 4.3 shows the average installation-labor cost by platform
type. The correlation between hull size and installation-labor cost was
0.88, strongly implying that the installation-labor cost increased for
larger ship types.

ISNS COMPOSE 3.0 Software. The average labor cost to install the
ISNS COMPOSE upgrade did not vary statistically across the different hull types, with one exception: The average labor cost to install the
upgrade on CVNs was much higher than that for any other hull type.
Figure 4.4 shows the average installation-labor cost by platform type.

EBEMs for WSC-6 Variants. The average labor cost to install the
EBEM upgrade reflected a different trend than those exhibited by the
upgrades just described. Here, the average cost decreased as the size of
the ship grew, a trend shown in Figure 4.5. This suggests that the instal-

Figure 4.3
Average Labor Cost to Install the ISNS LAN GIG-E Upgrade, by Hull Type

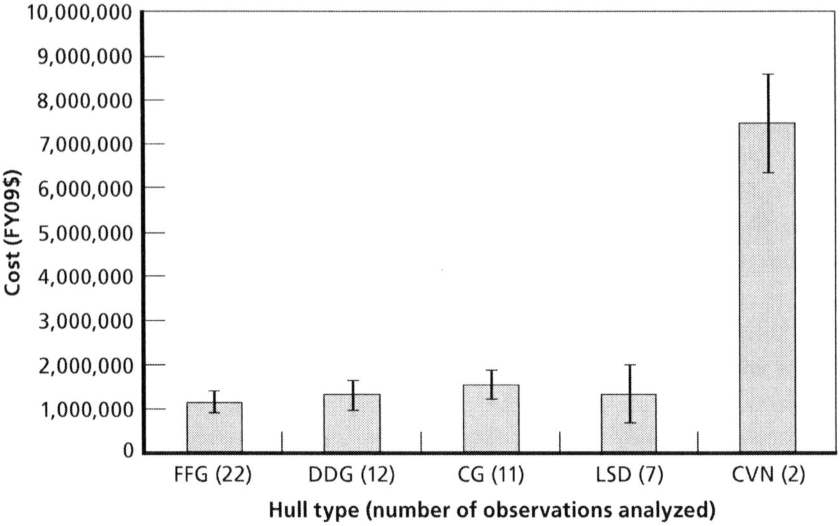

NOTE: The vertical line over each bar shows one standard deviation above and below the mean.
RAND MG907-4.3

lations might have been more difficult on the smaller ships because of the tighter space; or, it could be that installation on the smaller ships was more likely to occur at a more expensive location.

There was only one data point for EBEM installations on the LCC, so it was not possible to determine the statistical significance of the apparent separation. The difference between the average installation-labor cost for DDGs and those for CGs and LPDs was statistically significant. The CGs and the LPDs were very close in terms of mean labor costs to install the EBEM upgrade.

SSEE Increment E. The average labor cost to install the SSEE Increment E upgrade significantly increased as the size of the ship grew. The relative cost variability also increased with ship type. Figure 4.6 shows the installation-labor cost by platform type.

The analysis described in the preceding six sections suggests that the size of the platform typically had a direct impact on installation-labor costs. In the next section, we look at installations within each

Figure 4.4
Average Labor Cost to Install the ISNS COMPOSE 3.0 Software Upgrade, by Hull Type

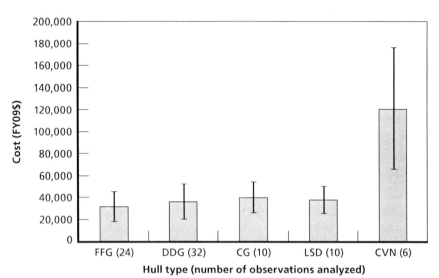

NOTE: The vertical line over each bar shows one standard deviation above and below the mean.
RAND *MG907-4.4*

ship type to identify other variables that may have influenced these labor costs. We focus on the age of the platform and the upgrade-installation location (i.e., Atlantic or Pacific coast).

Variations Within Ship Type

Our ability to conduct statistical analyses of the impact of ship age (as measured by start date minus commissioning date) and installation coast (Atlantic or Pacific) on installation-labor costs was limited by the small sample sizes available for some of the specific upgrades on some of the specific platform types.

Table 4.6 displays, by upgrade type and hull type, the results of regressions of installation-labor cost on ship age and installation coast. For the coast variable, we assigned the Atlantic Coast a value of zero and the Pacific Coast a value of one. The table reports the coefficients of age when such coefficients are statistically different from zero. The table

Figure 4.5
Average Labor Cost to Install the EBEMs for WSC-6 Variants Upgrade, by Hull Type

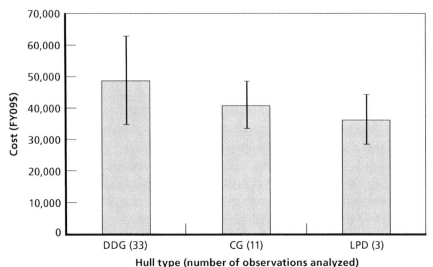

NOTE: The vertical line over each bar shows one standard deviation above and below the mean.
RAND *MG907-4.5*

also displays the labor-cost differences between the installation coasts when such differences are statistically significant; note that the difference is negative if installation was more expensive on the Atlantic Coast than on the Pacific Coast, and vice versa. In some cases, there were not enough observations from both coasts to allow us to perform a meaningful regression. In such instances, we omitted the installation coast from the regression. The mean cost is provided to put the coefficients in context. We also report the coefficient of determination—known as R^2—to indicate how much of the variability in cost is explained by the variables; the closer the number is to one, the more accurate the model.

There are several observations to be made from the results shown in Table 4.6. The primary observation is that **ship age and installation coast typically do not explain very much of the variation in installation-labor costs**. The two variables explained over 70 percent of the

Figure 4.6
Average Installation-Labor Cost of the SSEE Increment E Upgrade, by Hull
Type

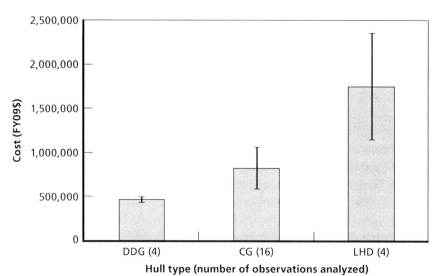

NOTE: The vertical line over each bar shows one standard deviation above and
below the mean.
RAND *MG907-4.6*

variation in cost in only two cases: the ISNS LAN GIG-E upgrade
on DDGs and the ISNS COMPOSE upgrade on LSDs. Second, the
age coefficient was only statistically different from zero in three of
the cases: ISNS Embarkable Drops on FFGs, ISNS LAN GIG-E on
DDGs, and SSEE Increment E on CGs. In these three cases, the implications of ship age were mixed. Upgrades on older ships cost more in
two cases, but the ISNS Embarkable Drops upgrade cost less on FFGs.
Finally, installation coast did have an impact in several cases, but the
implications were again mixed. In two cases—ISNS LAN GIG-E for
DDGs and FFGs—installations on the Atlantic Coast were much more
costly than the same installations performed on the Pacific Coast. In
three other cases—ISNS COMPOSE on DDGs, FFGs, and LSDs—
installations performed on the Atlantic Coast were less expensive than
those performed on the Pacific Coast.

Table 4.6
Regression of Installation-Labor Cost on Ship Age and Installation Coast, by Upgrade Type and Hull Type

Upgrade Type	Hull Type	Number of Installations Examined	Mean Cost (FY09$)	Age Coefficient ($ per year of age)	Cost Difference Between Installation Coasts (Pacific minus Atlantic, FY09$)	R^2
GCCS-M GENSER 4.X (V) 1–4	DDG	17	187,724	0	0	0.11
ISNS Embarkable Drops	FFG	13	58,438	−14,464	N/A	0.23
ISNS LAN GIG-E	CG	11	1,557,974	0	0	0.10
ISNS LAN GIG-E	DDG	12	1,303,677	84,786	−395,465	0.76
ISNS LAN GIG-E	FFG	22	1,137,324	0	−336,704	0.40
ISNS COMPOSE 3.0 Software	CG	10	39,887	0	N/A	0.01
ISNS COMPOSE 3.0 Software	DDG	32	36,198	0	15,582	0.25
ISNS COMPOSE 3.0 Software	FFG	24	31,437	0	12,607	0.20
ISNS COMPOSE 3.0 Software	LSD	10	37,882	0	19,991	0.74
EBEMs for WSC-6 Variants	CG	33	40,940	0	0	0.09
EBEMs for WSC-6 Variants	DDG	11	48,815	0	0	0.04
SSEE Increment E	CG	16	821,735	37,743	0	0.23

Obviously, our analysis was not able to capture a number of other factors that may have contributed to the large variability of installation-labor costs for the various upgrades we examined. Two such variables are the organization that installed the upgrade and whether the installation was accomplished at a private shipyard, at a public shipyard, or at an operating base. This last variable was often mentioned in interviews as a contributor to cost variability.

Cost Improvement Occurred with Some of the Upgrades

Cost improvement is a manufacturing observation that efficiency improves (and results in lower costs) as an activity is repeated multiple times. For example, doing something for the fourth time should cost less than each of the prior three times. Such efficiency gains are typically seen in assembly-line production, where multiple, identical items are produced and there is a substantial labor component. Our analyses suggest that **any cost-improvement effects in the upgrades were weak at best, but the results are not conclusive.**

Using linear regression to measure the relationship between the start date and the natural logarithm (Ln) of the installation-labor cost, we found both decreases and increases in installation-labor costs for successive installations for the different upgrades we analyzed.[12] The degree of learning suggested by the trend line of the coefficient of cost as a function of age is shown in Table 4.7. The table also shows whether the coefficient was statistically different from zero. The coefficient of the start date indicates the size of the learning effect, and the significance indicates whether the start date had a statistically significant impact on the change in the cost. If the coefficient is negative and significant, this provides evidence that learning effects may have reduced the labor cost of installation over time—an improvement. Alternatively, a positive

---

[12] The Ln of cost was used instead of actual cost because the model for cost improvement is an exponential trend. This transformation also mitigates the influence of outliers. The linear model used to find cost improvement was $Ln(Cost) = a*Start\ Date + b$. In this model, a is the coefficient of the start date, and b is the intercept. We also tested to see whether the coefficient was statistically different from zero.

Table 4.7
Cost-Improvement Statistics

Upgrade Type	Number of Installations Examined	Coefficient of Start Date	Significantly Different from Zero?
GCCS-M GENSER 4.X (V) 1–4	31	0.204	No
ISNS Embarkable Drops	34	−0.265	Yes
ISNS LAN GIG-E	54	0.110	No
ISNS COMPOSE 3.0 Software	82	0.332	Yes
EBEMs for WSC-6 Variants	48	−0.146	Yes
SSEE Increment E	24	0.219	Yes

coefficient indicates that there were inflationary effects or changes to the process that caused it to be more expensive.

The coefficients for the ISNS Embarkable Drops and EBEMs upgrades are negative and statistically significant, indicating that some cost improvement likely occurred for these two upgrades. The coefficients for the GCCS-M and ISNS LAN GIG-E upgrades are not statistically different from zero with any real confidence, which prevented us from making any conclusions about the impact of cost-improvement effects. Both the ISNS COMPOSE 3.0 Software and SSEE Increment E upgrades have positive coefficients that are statistically different from zero with a high confidence. This indicates that, for these two upgrades, the installation-labor costs increased with time.

Figures 4.7 and 4.8 show plots of the cost (Ln) versus the start date for the SSEE Increment E and ISNS Embarkable Drops upgrades, respectively. These figures illustrate the differing cost trends associated with these upgrades. The installation-labor costs associated with SSEE Increment E appeared to be increasing over time, while the installation-labor

Figure 4.7
Ln(Cost) Versus Start Date for SSEE Increment E

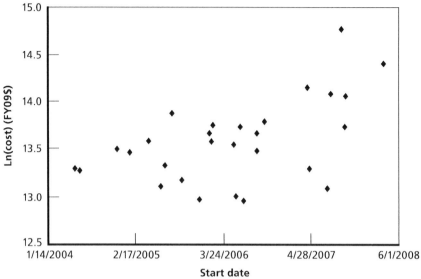

Figure 4.8
Ln(Cost) Versus Start Date for ISNS Embarkable Drops

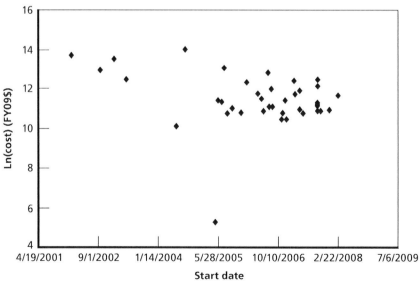

costs associated with the ISNS Embarkable Drops upgrade seemed to be declining slightly. Cost improvement is one possible reason why the ISNS Embarkable Drops costs were slightly declining.

Perhaps it is no surprise that we found inconsistent cost-improvement results across these C4I installations. Most of these installations took place at different locations, were performed by different contractors, and occurred on ships with varying configurations. Furthermore, some installations were concurrent with other installation work, which may have created work interference.

CANES Early Adopters

The CANES program is a computing-architecture initiative designed by the Navy to develop a single C4I computing environment for ships.[13] This initiative is anticipated to reduce both life-cycle costs and the effects of obsolescence on future ships. To date, no complete CANES installations have been performed; therefore, to determine the potential effect of the CANES effort, it was necessary to compare the installation-labor cost of upgrades to the first ships scheduled to get the CANES upgrade with the installation-labor cost of upgrades to other ships of the same hull type. We compared every installation on the CANES early adopters with identical installations on hulls of the same type. In each case, the deviation from the average installation-labor cost was determined, aggregated, and tested for significance.

Overall, it appears to have been more expensive to upgrade the ships selected for the CANES program than to upgrade non-CANES ships of the same type. The detailed results we list immediately below can be compared to future upgrade costs to determine whether CANES results in any savings. For instance, if future installations on the USS *Lincoln* cost the same on average as installations on other CVNs, does this imply that CANES had no effect? No, because installations on the USS *Lincoln* typically cost more than those on other CVNs, and this could be evidence that CANES lowered the installation cost. The results by ship are as follows:

---

[13] Turner, 2007.

- USS *Cape St. George* (CG-71): There did not appear to be a systemic difference between the labor cost of installations on the USS *Cape St. George* and those on other CGs.
- USS *Abraham Lincoln* (CVN-72): The labor cost for installations on the USS *Abraham Lincoln* was, on average, 49 percent of one standard deviation more than the mean cost of such installations on other CVNs. If the average installation-labor cost for a specific CVN installation type was $100,000, and the standard deviation was $50,000, then the expected installation-labor cost for the USS *Abraham Lincoln* is $124,500. This difference was statistically significant to the level of p = 0.99. The net effect of this difference was that the data set showed more than $24.5 million in installation-labor costs for the USS *Abraham Lincoln*—nearly $4 million more than any other CVN. However, this higher cost could be related to the fact that most of the USS *Abraham Lincoln*'s installations occurred during an availability at a public shipyard rather than at the operating base. There were no discernable patterns in cost differences between the installation types.
- USS *Momsen* (DDG-92): The labor cost of installations on the USS *Momsen* was generally about 16 percent of one standard deviation more than the mean labor cost of installations on similar ships. This difference was not statistically significant to the level of p = 0.90.
- USS *Russell* (DDG-59): The labor cost of installations on the USS *Russell* was generally about 49 percent of one standard deviation more than the mean labor cost of installations on similar ships. This difference was statistically significant to the level of p = 0.99.
- USS *Shoup* (DDG-86): The labor cost of installations on the USS *Shoup* was, on average, about 48 percent of one standard deviation more than the mean labor cost of installations on similar ships. This difference was statistically significant to the level of p = 0.99.
- USS *Sterett* (DDG-104): There were no installations in the data set for this ship.

These results suggest that, for most upgrade types, installation-labor costs for upgrades to CANES early adopters tend to fall in the upper range. The CANES ships were frequently among the first of their hull type to receive upgrades, which appears to have modestly contributed to the magnitude of the differences. Given the limited nature of the data set, it is difficult to say whether future CANES installs will be more costly in general, whether this cost premium was an additional cost to retrofit an existing hull, or whether this premium resulted from doing the work for the first time (and subsequent installations will be less costly). Also, this cost premium does not account for differences in the cost of hardware and other aspects of the life cycle that could offset the premium.

Estimate Accuracy

In the previous sections, we explored the drivers and variability of *actual* installation-labor costs. This section focuses on the accuracy of PMW's *estimates* of installation-labor costs. In other words, were the estimates close to the actual values? We define *estimate error*—generally referred to as *error* in the rest of this section—as the actual cost (or estimate at completion, if actual was not available) minus the PMW estimate.[14] This value is sometimes more commonly referred to as *cost growth*. As in our analysis of cost drivers, we used statistical testing (i.e., t-tests and ANOVA) to determine the significance of various error factors.

We used several different error metrics when comparing a project's installation-labor cost estimates with its actual installation-labor cost. Four of the formulas used to calculate error and bias metrics are presented below:[15]

---

[14] The data set frequently contained multiple values for the installation cost, and we used the actual installation cost from the most recent source. When information on neither the actual cost nor the estimate at completion was available, we omitted the data point from the analysis.

[15] In these formulas, n is the number of observations, x_i is the *ith* observation, p_i is the predicted value for x_i, and $\bar{x}$ is the mean value of the observations.

- mean absolute error: $\dfrac{\sum\limits_{i}\left|x_i - p_i\right|}{n}$

- relative absolute error: $\dfrac{\sum\limits_{i}\left|x_i - p_i\right|}{\sum\limits_{i}\left|x_i - \bar{x}\right|}$

- root mean squared error: $\sqrt{\dfrac{\sum\limits_{i}\left(x_i - p_i\right)^2}{n}}$

- mean bias: $\dfrac{\sum\limits_{i}\left(x_i - p_i\right)}{n}$.

A fifth metric, the correlation between the predicted installation-labor cost and the actual installation-labor cost, is also a good measure of estimate accuracy. This measure is independent of scale, so it is possible to compare its value across any groupings. Correlation values range between one and negative one. A value of one indicates a perfect correlation between the two variables. The smaller the correlation's magnitude, the less related the two variables are. A negative value indicates that the variables move in opposite directions (e.g., when one variable increases, the other tends to decrease). The correlation ignores systematic biases in the data; therefore, if every installation-labor cost estimate were exactly $1,000 below the actual installation-labor cost, the correlation would be one despite the bias to underestimate the cost.

Mean absolute error and root mean squared error are two metrics of the difference between actual installation-labor cost and estimated installation-labor cost. The mean absolute deviation is much less sensitive to outliers than is the root mean squared error. Thus, a comparison of these two metrics can provide some indication of the impact that cost outliers have on estimates. If these two error metrics are wildly different, most of the error is the result of a few outliers.

Relative absolute error is a normalization of the mean absolute error. Because this metric is normalized, it can be used to compare different groups (e.g., installation types). It can be thought of as a measure of how much more accurate an estimate is compared with the average value of a sample. The lower this number, the better. For budgeting, most estimating systems aim for a value in the range of 10–15 percent.

Mean bias measures the systematic error of the estimates. This metric is the sum of the difference between the actual installation-labor cost and the estimated installation-labor cost. If the predictions tend to underestimate the installation-labor cost of an upgrade installation, the bias will be positive, and vice versa. In the remainder of this monograph, the terms *bias* and *mean bias* are used interchangeably.

We found 3,707 underestimates, 4,947 overestimates, and 608 exact estimates. Figure 4.9 is a log-scale scatter-plot of the actual installation-labor cost divided by the estimated installation-labor cost. An estimating system that perfectly predicts cost would result in points that form a straight line with a value of one on the y-axis. The figure shows that there is a great deal of variability in the accuracy of the estimates. The installation-labor costs of installations with very small actual installation-labor costs are quite frequently drastically overestimated. Additionally, it appears that the relative variability decreases as the installation-labor cost increases (i.e., the spread of points decreases for higher values). It also appears that, for installations with very large actual installation-labor costs, there is a slight tendency to underestimate the costs.

Figure 4.10, a histogram of actual installation-labor costs divided by estimated installation-labor costs, provides an idea of the relative magnitude of error. A value of one implies a perfect estimate. A value of greater than one is an overrun (i.e., the estimated cost was lower than the actual cost), and a value of less than one is an underrun (i.e., the estimated cost was greater than the actual cost). Note that the distribution before and after one on the x-axis is not quite symmetric (symmetry would be desirable) and that there is a larger count for values less than one. These two features imply that **the system has some bias toward overestimating actual costs and exhibits a great deal of variability**.

Figure 4.9
Actual Installation-Labor Costs Divided by Estimated Installation-Labor
Costs, by Actual Cost

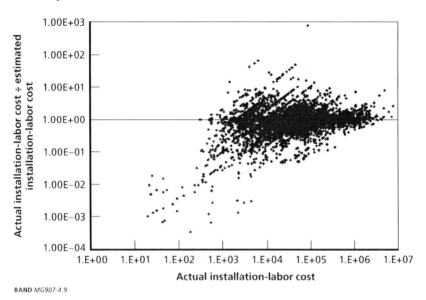

Figure 4.10
Magnitude of Estimate Errors

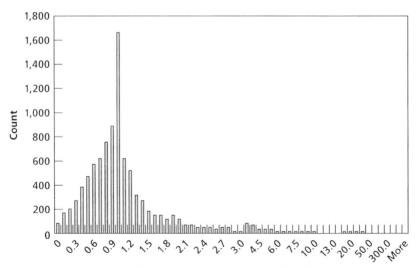

Statistically, the estimated installation-labor costs have a correlation of 0.90 with the actual installation-labor costs. This implies that the estimated costs generally increase when the actual costs increase, and vice versa. The mean absolute error is more than $47,000, implying that, on average, the predictions are off by that amount. The mean bias of the errors was an overestimate of $7,112. About one seventh ($7,112 ÷ $47,713) of the total mean absolute error was a direct result of that bias. The bias can provide some indication about what causes these estimating errors (i.e., one-seventh of the errors are due to a systematic tendency to overestimate). The relative absolute error is sufficiently high that the effects of any cost-reduction measures will be difficult to observe. The ideal upper bound for relative absolute error depends on the context. Again, a target of below 15 percent for a budget-value estimate is desirable, with lower values being preferable. Given the high variability discussed earlier in this chapter, this target could be hard to reach. Table 4.9 lists the estimate error statistics by hull type. The estimates for the CVN, FFG, LCC, and SSN hull types stand out as particularly poorly correlated with actual cost.

For several hull types, systematic biases had a significant impact on the accuracy of estimates. More than half the mean absolute error for LSDs was the result of biases. Likewise, more than 30 percent of the mean absolute error for LHAs and LCCs was directly attributable to systematic overestimates. Our analysis shows that there is strong evidence that outliers (i.e., upgrades that cost significantly more or less than estimated) drive these errors.

Analysis by Hull Type

In this section, we look more closely into the hull types whose cost estimates poorly matched actual costs.

CVNs. The estimated costs of upgrades to CVNs were the least accurate by most measures. Our analysis indicates that outliers drove the error more than a systematic bias did. Most of these errors were evident in upgrades to networks and communications; the worst errors in absolute terms were associated with ISNS LAN GIG-E variants. The errors in the estimates of network-upgrade labor costs were significantly higher than those found in other upgrade types (p = 0.99), and

Table 4.9
Accuracy of Estimates by Hull Type

Hull Type	Correlation	Mean Absolute Error (FY09$)	Relative Absolute Error (percentage)	Root Mean Squared Error (FY09$)	Mean Bias (FY09$)
CG	0.9134	43,076	26.28	136,825	−8,154
CVN	0.8411	72,693	40.19	291,916	3,344
DDG	0.9578	26,934	18.90	109,025	−4,309
FFG	0.8440	34,107	40.66	104,784	−7,933
LCC	0.8550	67,973	35.48	227,494	−19,211
LHA	0.9443	55,167	29.58	134,612	−19,989
LHD	0.9535	56,943	28.20	159,092	1,827
LPD	0.9194	44,086	32.20	121,716	−8,529
LSD	0.9060	50,046	32.56	157,545	−26,566
SSN	0.8471	51,434	27.51	196,588	−9,186
Total	0.8969	47,713	28.29	168,777	−7,112

the errors found in the communications-upgrade estimates were higher than those found in the remaining upgrade types (p = 0.90). There was no statistically significant bias for communications or network upgrades, but there was significant evidence for bias in C2 upgrades (p = 0.99). These upgrades tend to overestimate the cost for installation by around $20,000.

There was no statistically significant bias in the overall errors, and there was no statistically significant difference in errors based on the installation coast. There were no statistically significant differences based on availability type, but there could be a small bias toward underestimating the installation-labor cost of WOO upgrades.

The correlation between start date and absolute error was −0.11. This value is very small, but it could imply that the estimates were improving over time.

FFGs. There was evidence that installation-labor costs were being systematically overestimated for this hull type. Actual installation-labor costs over the data set amounted to about $59 million, whereas estimated installation-labor costs were about $65 million—a net difference of nearly $6 million. Much of this error occurred in installations whose actual installation-labor cost was less than $1,000. For example, the installation-labor cost estimate for ISNS Embarkable Drops for the USS *Curtis* was $618,898, and the actual installation-labor cost was $203. The 20 largest errors occurred in either the ISNS LAN GIG-E or ISNS Embarkable Drops upgrades. Because of the magnitude of many of these differences, it seems likely that some of this error was the result of data-entry mistakes.

In terms of magnitude of error, the estimated installation-labor cost of installations done on the Atlantic Coast was statistically worse than that of installations performed on the Pacific Coast (p = 0.99). The estimates for upgrades performed on the Atlantic Coast also contained a statistically significant bias (p = 0.90) toward overestimating installation-labor cost. The bias for upgrades performed on the Pacific Coast was not statistically significant.

There was no evidence of systematic bias in the estimates of the installation-labor costs of the WOO upgrades, but there was evidence of installation-labor cost overestimates for the selected restricted availability upgrades (p = 0.95) and other types (p = 0.99). Additionally, the selected restricted availability–related estimates featured the highest mean absolute error. Next were the WOO upgrades and then the other types. The separations between these upgrade types were statistically significant to p = 0.95.

There was no evidence of improvement in the estimates over time.

LCCs. The data set contains two LCCs: the USS *Blue Ridge* and the USS *Mount Whitney.* The differences in real absolute error and bias between the two hulls were large (an error of $50,000 and an underestimation bias of $13,000 for the USS *Blue Ridge* compared with an error of $92,000 and an underestimation bias of $28,000 for the USS *Mount Whitney*), but, because of the large variability, they were not statistically significant.

SSNs. For the SSN hull type, there was a statistically significant bias (p = 0.90) toward overestimating installation-labor costs. The total of the estimates exceeded the total of the actual costs by more than $15 million out of a total of nearly $273 million spent. The largest single source of error (most commonly, overestimates) in the estimate was the Install TIDS Phase II BSY-1. Additionally, the estimates for the Install GCCS-M 4.X upgrade overestimated the installation-labor cost by approximately $450,000–$550,000 in every case. This overestimating error could have been due to either accounting issues or scope reductions because, although the actual installation-labor cost was generally around $25,000, the estimated installation-labor cost was closer to $500,000.

The estimates for installation-labor costs for any docking upgrade (e.g., a docking selected restricted availability or a docking planned incremental availability) were significantly less accurate (p = 0.95) than those for other types. There was no statistical difference between the WOO upgrades and any upgrade type other than the docking upgrades.

There was evidence in the PMW 150–supplied data of a systematic tendency to overestimate of the installation-labor costs. Estimates for C2 and network upgrades were statistically less accurate (p = 0.95) than those for other types of upgrades. There was no evidence that the installation coast affected the accuracy of the installation-labor cost estimate. There was no significant evidence that estimates improved with time.

Analysis by Upgrade Type

In this section, we look into the accuracy of the estimates for the six specific upgrades we examined earlier in this chapter.

For the ISNS Embarkable Drops upgrade type, all of the 43 estimates but one overestimated the actual installation-labor cost. Installation coast did not appear to affect the error, but there was a positive correlation (0.39) between start date and absolute error. This implies that the predictions were getting worse over time. There was a statistically significant overestimation bias (p = 0.95).

For the ISNS LAN GIG-E upgrade type, the estimates appeared to be getting slightly worse over time, and the correlation between start date and cost-estimate error was 0.20. This correlation did not appear to be a result of variation in hull types. There was a bias toward under-estimating the labor cost of installation (by about $320,000, on average), and there were about twice as many underestimates as overestimates. All of the WOO installations were underestimated. Upgrades performed on the Atlantic Coast were significantly underestimated ($p = 0.99$), and there was no statistically significant bias in estimates for upgrades performed on the Pacific Coast. However, there was statistically no difference between the coasts in terms of the total mean absolute error. Thus, estimates for the coasts were, overall, equally accurate, but the estimates for upgrades performed on the Atlantic Coast were significantly biased.

For the ISNS COMPOSE 3.0 Software upgrade type, the estimates were essentially independent of start date and installation coast. There did not appear to be any patterns in the error due to availability type. Estimates for upgrades performed on the Atlantic Coast were significantly biased ($p = 0.99$), and there was no statistically significant bias in estimates for upgrades performed on the Pacific Coast. However, there was statistically no difference between the coasts in terms of the mean absolute error. For this upgrade type, estimates for the coasts were, overall, equally accurate, but the estimates for upgrades performed on the Atlantic Coast tended to overestimate the cost.

For the GCCS-M GENSER 4.X (V) 1–4 upgrade type, the errors and the start time had a correlation coefficient of –0.21, implying that the estimates were getting better over time. There were no other discernible patterns in the errors for this upgrade type.

For the EBEMs for WSC-6 Variants upgrade type, estimates were systematically lower ($p = 0.99$) than the actual installation-labor cost by an average of about $15,500. There did appear to be a weak improvement with the estimates over time because the correlation between error and start time was –0.12. There was not a statistically significant difference between the biases evident in estimates for upgrades performed on either of the two coasts, but estimates for upgrades performed on the Pacific Coast were statistically significantly less accurate, in terms of mean absolute error, than those calculated for upgrades performed on the Atlantic Coast ($p = 0.95$). Different availability types did not appear to lead to statistically significant differences.

For the SSEE Increment E upgrade, estimates displayed a statistically significant underestimation bias (p = 0.95) of over $220,000, on average. There were 22 overestimates and just eight underestimates. There was not a statistically significant difference in terms of installation coast or availability type.

Summary of Observations

In this chapter, we analyzed historical installation-labor–cost data to address the following questions:

- What factors affect the labor cost of installing C4I upgrades?
- Do C4I-upgrade installation-labor costs decrease for successive installations of a given upgrade (i.e., is there learning)?
- Are installation-labor costs different for CANES early adopters compared with other ships in their class?
- How well are C4I installation-labor costs estimated?

For the specific installations we analyzed, the level of variability in actual installation-labor cost, even within a ship class, was quite high: The high-to-low value was an order of magnitude different. Our analysis suggests that for installations of the SSEE Increment E, EBEMs, ISNS, and GCCS-M upgrades, ship size frequently had a significant impact on the labor cost of installation. However, within a given class of ships, the age of the ship at the time of the upgrade and the installation coast typically explained very little of the variation in installation-labor cost. Even when ship age and installation coast did explain part of the variability in installation-labor cost, the results were mixed. In some cases, labor-installation costs were greater for older ships; in other cases, the opposite was true. In some cases, installations accomplished on the Pacific Coast were more expensive than those performed on the Atlantic Coast; in other cases, we saw the opposite effect. In some cases, the installation coast had a significant effect on installation-labor cost.

We did see some evidence of learning. That is, in some cases, the labor cost of installing an upgrade decreased for successive installations.

But, there was at least one case in which negative learning occurred. The CANES early adopters tended to accrue higher installation-labor costs than similar ships. Finally, cost estimates tended to overestimate the labor cost of upgrades, and the relative error was quite high, particularly for CVNs, FFGs, and SSNs. Many of the factors that were found to influence installation-labor cost variability, such as hull type, hull age, and installation coast, also affected estimation accuracy.

In addition to these observations, our analysis of the data uncovered several interesting findings with potential policy implications. First, we note that three additional pieces of information might be valuable for understanding cost variations if tracked and collected as part of the Space and Naval Warfare Systems Center/PEO Integrated Data Environment and Repository database: installation hours, the organization that performed the installation, and in what type of facility the installation was performed (i.e., public shipyard, private shipyard, or operating base).

Second, creating after-action reports after upgrades are completed could help estimators improve the quality of their estimates in the future. To improve the installation-labor cost predictions, **after-action reports should be created when the estimates of installation-labor cost have proven dramatically off the mark**. Over time, this practice should reduce the systematic misestimating of installation-labor costs. For example, installing GCCS-M GENSER 4.X (V) 1–4 for SSNs cost around $25,000 in labor, but the estimates were closer to $500,000. The estimate for the first installation was found to be significantly off, and an investigation into the reasons why could have prevented the labor cost of the nine upgrades that followed from being overestimated (by a total of nearly $5 million). After-action reports should be collected by cost analysts and periodically mined for relevant information. This practice could be especially useful for identifying particular strategies that particular sites employ that result in *real* installation-labor cost reductions (i.e., reductions that are not simply the result of estimate errors or accounting or reporting problems).

CHAPTER FIVE

Conclusions and Recommendations

Navy organizations, especially PEO C4I, and ship designers and ship-builders have recognized the high costs associated with constantly upgrading C4I systems, and they have therefore adopted numerous design features to make C4I upgrades easier and less expensive. These design features have helped make C4I upgrades on ships less costly and less time-consuming. However, these features have largely been applied to individual classes of ships and are not yet standard across the fleet. Moreover, some features may not be appropriate for all ships. Consequently, although the average labor cost to install C4I upgrades on in-service ships has decreased over the last decade, the labor to install C4I upgrades still costs the Navy more than $100 million annually.

Our findings indicate that changing how C4I systems are managed during design and construction is as important as adopting innovative design features to controlling C4I-upgrade costs. If the Navy does not seek innovative program-management approaches as well as innovative design features, funding C4I-system upgrades will remain a problem for the Navy. Our results also show that installation-labor costs for C4I upgrades vary significantly: There is often a difference of an order of magnitude or more between the lowest and highest labor costs of installing any specific upgrade. Our analysis of historical data suggests that ship size often affects the labor cost of upgrading a specific system. However, ship age and installation coast typically had little impact on the labor cost of installing a specific upgrade across a class of ships.

Finally, the Navy needs to be able to better predict labor costs associated with installing C4I systems; otherwise, it will not be able to determine whether planned or actual improvements have achieved the desired results. A number of recommendations, discussed below, flow from these conclusions.

Designing and Building Ships to Reduce C4I-Upgrade Costs

From the perspective of designing and building naval ships to facilitate C4I upgrades, the Navy should expand or better emphasize the following initiatives:

- **Ensuring that adequate design margins for power, cooling, and space are incorporated into the design of a ship and that adequate margins are sustained during the operational life of a naval ship.** Adequate margins are often set early in the design process, but they are typically reduced during the design and construction phases when costs increase. New C4I capabilities are always becoming available, and fleet and ship commanders always want these new capabilities and technologies. Without adequate power, cooling, and space to accommodate these new capabilities, C4I-upgrade costs grow.
- **Including adequate access paths when designing a ship, especially a surface combatant.** Upgrade costs can increase if it is difficult to remove old equipment and install new equipment. Designers have incorporated wide passageways and easy access to various equipment into their designs for submarines (especially the *Ohio* and *Virginia* classes) and some amphibious ships.
- **Using standard racks and fixtures and flexible infrastructures to help reduce the amount of hot work required to remove old fixtures and install new ones when upgrading C4I systems.** The use of standard racks in submarines and the DDG-51 class of ships has successfully reduced upgrade costs. New equipment can be mounted and tested before installation on the ship, and

new racks with the upgraded equipment can quickly replace the racks that host the old equipment. The *Ford* class of aircraft carriers is adopting a flexible infrastructure for many of the spaces on the ship, which will allow spaces to be quickly reconfigured with no or minimal hot work and without the need to run new power cables or cooling ducts.

Each of these approaches is currently in use, but to what extent depends on the ship class. Sometimes, these approaches are the first to be cut when affordability issues arise.

A few issues require further analysis and resolution:

- **GFE versus CFE, federated versus integrated systems, and SOAs versus OAs.** There are advantages and disadvantages to both sides of each pair. When analyzing these options, the Navy must recognize that technologies that work well in the commercial world may not be best for naval ships.
- **The SHIPMAIN process for C4I upgrades.** Interviewees often described the SHIPMAIN process as a hurdle to installing C4I upgrades in a timely manner. However, others noted that waivers or accelerated processes are available in certain situations.

Managing Ship Programs to Reduce C4I-Upgrade Costs

Effectively managing C4I systems during ship design and construction is as important as the features designed into the ships for controlling C4I-upgrade costs. It takes several years to build a naval combatant, and the C4I systems can refresh multiple times during that period. Program managers must ensure that the most-up-to-date technologies are incorporated in the ship when it is delivered. This is difficult because ship designers and shipbuilders want all the specific details and equipment to be defined early in the construction process. Also, some C4I systems must be installed early in the build process to support the ship's crew or to facilitate testing. To effectively manage ship programs when installing new C4I capabilities, the Navy should focus on the following areas:

- **Technology insertion.** This process eliminates the need to rip out "old" technologies and install new ones as soon as the ship is delivered to the Navy. Many programs have adopted initiatives (termed *design budget, technology insertion,* or *turnkey*) to help deliver a ship with the most-current C4I technologies. These initiatives typically define the space, power, and cooling requirements for the C4I systems without specifying the exact systems that will be installed on the ship. Shipbuilders construct the C4I spaces with the specified requirements, and the Navy then provides the equipment itself just before the ship is delivered.
- **Configuration management.** One difficulty associated with upgrading in-service ships is that ships within a single class can exhibit various configurations. These various configurations mean that each installation of a specific upgrade is unique. This raises the cost of planning the upgrade and prohibits any learning that could otherwise have reduced installation costs across the ships in a class. Program managers must work to control the configurations of the various ships within a given class. Submarine programs appear to do this effectively: Typically, all submarines in a given class maintain the same configuration.

Steps to Address the Variability in Upgrade Costs

The average labor cost of a C4I upgrade is over $150,000 per installation. Furthermore, the installation-labor costs for C4I upgrades vary significantly: As noted earlier in this chapter, there is often a difference of an order of magnitude or more between the lowest and highest labor costs of installing any specific upgrade. Analyses of six specific types of upgrades seemed to identify some of the causes of the large variability, and this may provide insight to Navy planners as they budget and plan capability upgrades. However, because a large portion of the variability in installation-labor cost was not explained by the various factors we examined, we suggest in the remainder of this chapter a set of steps to support further analyses.

The Impact of Ship Type, Ship Age, and Installation Location
We arrived at the following conclusions:

- **Navy planners should anticipate higher installation-labor costs for larger ships.** The size of a ship (i.e., the class of the ship) typically had a significant impact on the labor cost of a given type of installation. The one exception was the EBEMs for WSC-6 Variants upgrade, where larger ships had lower installation-labor costs. It is not unreasonable that installation-labor costs for many types of upgrades increase as the size of the ship increases. Larger ships typically require more terminals and routers and that more cable and ducting be run through the ship.
- **The labor cost to install an upgrade was typically greater for older ships, but the results were mixed.** Our analysis found that labor for installing upgrades on older ships usually cost more, but the opposite proved true in the case of the ISNS Embarkable Drops upgrade.
- **Installation coast is not, at present, a meaningful variable for installation-labor cost estimates.** The Navy has insufficient data to determine whether installation-labor costs were more expensive for the Atlantic fleet compared to the Pacific fleet. In the few cases where there were sufficient data points for statistical analysis, the results were mixed: In some cases, installation-labor costs on the Atlantic Coast were higher; in other cases, installation-labor costs on the Pacific Coast were higher.
- **The Navy should add two variables—the installation facility (i.e., public shipyard, private shipyard, or operating base) and the team doing the installation—in future revisions to the database.** We could not examine the impact of these variables because the needed data were not included in the database. Interviewees suggested that labor costs were greater if the installation was accomplished at a shipyard, especially a public shipyard, rather than at the operating base. Also, interviewees suggested

that some installation teams performed better (and, therefore, cost less) because of their experience and preparation.

- **After-action reports would help manage installation-labor costs.** Currently, there is no record of the specifics of a given installation. After-action reports could provide insights into the factors that contributed to the total labor cost of an installation.

Reducing Costs for Successive Installations

In many industrial and manufacturing situations, costs decrease as an activity is accomplished more frequently. As noted in Chapter Four, this cost improvement for successive items or actions is often referred to as *learning.* Our analyses of the six different upgrade types suggest there was a decrease in installation-labor costs for two upgrades: EBEMs for WSC-6 Variants and ISNS Embarkable Drops. The installation-labor costs of these upgrades decreased as successive installations were accomplished. However, there was negative learning (i.e., an increase in installation-labor costs for successive upgrades) for the ISNS COMPOSE and SSEE Increment E upgrades. This finding was troublesome because it suggests that teams needed more hours to complete each successive installation of these upgrades—exactly the opposite of what one would expect.

Accuracy in Estimating Installation-Labor Costs

To improve the accuracy of cost estimation, after-action reports should be created whenever labor-cost estimates have proven dramatically off the mark. We used several metrics to assess the accuracy of the estimates of C4I-upgrade installation-labor costs. The mean absolute error across all the installations in the database was almost $48,000, suggesting that **installation-labor costs were typically overestimated.** Overestimating was a **particular problem for installations on larger ships,** such as aircraft carriers and amphibious ships. Also, overestimates were especially large for installations whose actual installation-labor costs were low. Overestimates are a problem because they tie up money that could be used for other projects.

Programs Managed by PEO C4I

PEO C4I supports naval forces through the provision and support of various C4I systems throughout the service. A large number of different systems are managed by PEO C4I. Included in Table A.1 are the programs currently under the program management of units within PEO C4I. The acquisition category (ACAT) defines the size of the program.[1] ACAT I programs are deemed Major Defense Acquisition Programs and are estimated to require either total research, development, test, and evaluation expenditures of more than FY96$355 million in a fiscal year or total procurement expenditures of more than FY96$2.135 billion. ACAT II programs are typically those that do not meet ACAT I criteria but are estimated to require either total research, development, test, and evaluation expenditures of more than FY96$140 million or total procurement expenditures of more than FY96$645 million. ACAT III, ACAT IV, and variants of the ACAT I and ACAT II programs are lower-cost programs.

PEO C4I is the program manager for six ACAT I programs, five ACAT II programs, and 45 ACAT III, IV, pre-ACAT, and Rapid Deployment Capability (RDC) programs. These are listed in Table A.1.

---

[1] See Department of the Navy, undated; Secretary of the Navy, 1996, for additional information on Navy programs and regulations governing major defense acquisition programs, respectively.

Table A.1
Programs Managed by PEO C4I

Name	Abbreviation	ACAT	Description
Consolidated Afloat Networks and Enterprise Services	CANES	I	CANES is the consolidation and enhancement of the requirements for five existing legacy-network programs. It is also a single support framework for all C4I applications that currently require dedicated infrastructure to operate delivered and managed legacy systems. These applications include ISNS, SCI Networks, and CENTRIXS-M. The CANES concept requires a technical and programmatic realignment of afloat infrastructure and services. CANES will take advantage of the new business model of OA, SOA, and rapid COTS insertion to bring fiscal savings to the Navy and operational agility to the warfighter.
Deployable Joint Command and Control	DJC2	I	DJC2 is a Secretary of Defense and Chairman of the Joint Chiefs of Staff priority transformation initiative. DJC2 will provide a standardized, integrated, rapidly deployable, modular, scalable, and reconfigurable JC2 and collaboration system to geographic Combatant Commanders to support en-route and initial-entry C2 and scale to support the full Joint Task Force Combat Operations Center.
Global Command and Control System–Maritime	GCCS-M	I	GCCS-M is the Navy's fielded C2 system. It provides the warfighter with a current status report of the vital positional information and data needed to make tactical decisions. The program is managed as an evolutionary acquisition system that facilitates rapid insertion of new functionality, technology, and COTS products.
Naval Tactical Command Support System	NTCSS	I	NTCSS is a multi-application program that provides standardized tactical-support information-systems capability to afloat, deploying, and shore-based Navy and Marine Corps activities. NTCSS incorporates aviation, surface and subsurface maintenance, supply, inventory, finance, and administration.
Navy Extremely High Frequency SATCOM Program	NESP	I	NESP provides secure, survivable, antijam, and low-probability-of-interception/detection communications terminals designed to provide connectivity to a variety of strategic and tactical C3I applications.

Table A.1—Continued

Name	Abbreviation	ACAT	Description
Navy Multiband Terminal	NMT	I	NMT is the fourth-generation MILSATCOM system for naval platforms. NMT will provide joint interoperable "core" and "hard-core" communications (via AEHF, Milstar, UFO, Polar, DSCS, and WGS satellites) and protected and wideband communications (Q/Ka/X [ship], Q/X [submarine], Q [shore], and GBS receive [ship/submarine]).
Command and Control Processor	C2P	II	Next-generation C2P provides a system capable of supporting critical data-link functions, including simultaneous processing of Link-11, Link-16, Link-22, Joint Range Extension, and High Throughput Link-16. NGC2P will be integrated into existing CDLMS system hardware and software and fielded primarily through SHIPALT and field changes.
Common Submarine Radio Room	CSRR	II	CSRR supports the Navy's evolving approach to network-centric–warfare IP-based secure communications. CSRR promotes commonality of equipment across the five submarine classes and leverages the *Virginia*-class ECS for the high data throughput required to support future operations. CSRR maximizes COTS and state-of-the-art technology in an OA to support technology insertion and technology refresh efforts.
Distributed Common Ground System–Navy	DCGS-N	II	DCGS-N provides the Navy's primary intelligence, surveillance, reconnaissance, and targeting support capability. Whether afloat (i.e., on a CVN, LHA, LHD, or LCC) or ashore (i.e., within an MHQ or a MOC), DCGS-N's tools are critical to both the operational commander's battlespace awareness and net-centric operations.
Integrated Shipboard Network System	ISNS	II	ISNS provides the shipboard network infrastructure for all existing C4I and business applications (more than 200 applications total), including GCSS-M, NTCSS, DMS/SMS, NSIPS, and the Navy Marine Corps Portal.
Super High Frequency Shipboard Terminal	AN/WSC-6(V)	II	This system provides reliable high-capacity (2.048 Mbps, E1) and interoperable worldwide communications for naval and joint warfighting and NATO interoperability via DSCS and WGS. The (V)9 version is also commercial–C-Band capable.

Table A.1—Continued

Name	Abbreviation	ACAT	Description
Automated Digital Network System	ADNS	III	ADNS Increment I provided a WAN router, allowing IP traffic from different enclaves to be combined and transmitted across a single radio-frequency path. Compatibility of traffic processed by the ADNS system was maintained across the unclassified, SECRET, SCI, joint, and coalition networks.
Battle Group Passive Horizon Extension System–Surface Terminal	BGPHES-ST	III	This signal-acquisition system extends a battle group's line-of-sight VHF/UHF radio horizon by using remote receivers in an airborne sensor payload and transferring SIGINT via SHF links to surface ships. The system provides local receivers for MF/HF/VHF/UHF signals of interest.
Commercial Wideband Satellite Program	CWSP	III	The AN/WSC-8(V)1/2 terminal provides a full duplex of E-1 (2.048 Mbps) to the fleet through a family of COTS/NDI SATCOM terminals and services. Products delivered include imagery, telemaintenance, telemedicine, SIPR/NIPR, secure phones, POTS, and VTC.
Common Data Link–Navy	CDL-N	III	The CDL-N system provides a wideband data link between Navy/joint airborne sensors and shipboard processors of national and tactical reconnaissance programs for real-time exploitation.
Communications at Speed and Depth Antenna System	CSD	III	This system provides a submerged communications capability to U.S. submarines. The CSD family of systems uses tethered buoys, a modified Buoyant Cable Antenna (to utilize the HF IP system), UHF SATCOM tethered buoys, and an improved RF-to-Acoustic Gateway Buoy. CSD systems will be installed and deployed from SSNs and SSGNs; there will be limited installations on SSBNs.
Cooperative Outboard Logistics Update	COBLU	III	COBLU is a joint U.S./UK project whose purpose is to update the existing outboard system (AN/SSQ-108) to provide the navies of the United States and the United Kingdom the capability to carry out comprehensive surface tactical information-warfare exploitation and electronic-warfare support measures.

Table A.1—Continued

Name	Abbreviation	ACAT	Description
Digital Modular Radio	DMR	III	DMR is a digital, modular, software-programmable, multichannel, multifunction, and multiband (2 MHz –2 GHz) radio system with embedded INFOSEC. DMR provides improvements for fleet radio requirements in the HF, VHF, and UHF bands and is interoperable and backward-compatible with legacy systems.
Global Broadcasting Service Shipboard Antenna System	GBS	III	This is a joint ACAT 1CD program. Due to differences in the Navy's component of the GBS system, the Navy decided to start a separate ACAT III program. However, the Navy has been procuring the Receive Broadcast Manager via a contract with the GBS Joint Program Office, with the Air Force serving as executive agent. This arrangement allows the Navy to gain economies of scale for large-quantity buys. Future procurements may occur through a separate contract.
High Frequency Radio Group	HFRG	III	HFRG is a solid-state broadband HF communication system consisting of transmit, receive, and control subsystems. Features include centralized remote control, automated BIT/BITE, rapid and reliable circuit reconfiguration, increased reliability, and reduced channel separation.
International Maritime Satellite Program	Inmarsat	III	Inmarsat B HSD provides (1) continuous, full-period, leased-channel-mode service for simultaneous processing of NIPRNET, SIPRNET, and Joint Worldwide Intelligence Communications System applications at speeds of up to 128 Kbps and (2) multiple telephone lines. This includes three official telephone lines, two of which are Secure Telephone Unit III–capable and one of which is fax-capable. There is also an unofficial crew telephone line for afloat personal telecommunications.

Table A.1—Continued

Name	Abbreviation	ACAT	Description
Joint Cross Domain Exchange	JCDX (OED)	III	The JCDX system supports command, control, and intelligence assessment, including I&W and power projection; maintains dynamic databases to support a common air, land, sea, and littoral battlefield picture using ground-force and maritime symbology; provides access to multiple communications networks for interforce compatibility and interoperability that support database sharing and data analysis; and supports Joint Task Force commanders, CINCs, service components, and subordinate units. JCDX will operate in a multilevel, secure DII COE to provide local and global networking for on-demand services and timely response to consumer requests for fused intelligence. JCDX supports joint Air Force, Army, Navy, Marine Corps, and Coast Guard operations with additional tasking to support counterterrorism, homeland security, counternarcotics, and allied coalition operations. JCDX directly contributes to the following ASD C3I CIO goals: (1) make information available on a network that people depend on and trust, (2) deny the enemy comparable advantages, and (3) exploit weaknesses.
Mini Demand-Assigned Multiple Access	Mini DAMA	III	Mini-DAMA is a communication system that supports the exchange of secure and nonsecure battlegroup coordination data, tactical data, and voice between baseband processing equipment over UHF SATCOM, 25/5 kHz DAMA, 25/5 kHz Non-DAMA, and UHF LOS. The system will include Mini DAMA, AN/WSC-3(V)6/7/11 LOS RADIOS, SSA, TD-1271, VICS, and HSFB.

Table A.1—Continued

Name	Abbreviation	ACAT	Description
Navigation Warfare Sea	NAVWAR SEA INC 1, 2, 3	III	The GAS-1 is a controlled reception pattern antenna that provides antijam nulling protection for GPS signals. It is a joint service product currently used by the Air Force and several allied countries. It has been identified as an initial replacement for the U.S. Navy's existing Fixed Reception Pattern Antenna on all Navy ships and on several aircraft platforms. The GAS-1 is one of several products that will be integrated into naval platforms as part of the Navy's Navigation Warfare Program. This program will implement antijam protection and other GPS modernization enhancements to ensure the continued viability of GPS-signal availability in terms of position, timing, and accuracy to support the Navy's warfighting capability.
Sensitive Compartmented Information Networks	SCI	III	This network provides network enterprise services for the afloat SCI Enclave, secure WAN IP access to ship and shore national Web sites, and SIGINT and intelligence databases. Its primary role is to transport and disseminate special intelligence tactical information and exchange, in near–real time, time-sensitive intelligence and tactical cryptologic sensor data, voice, video, and SCI record-message traffic between ships and between ships and the shore. SCI Networks is the SI FORCEnet enabler.
Shipboard Single Channel Ground and Airborne Radio System	SINCGARS	III	This system provides secure, antijam VHF voice and data communications. The SINCGARS System Improvement Program (SIP) and the SINCGARS Advanced System Improvement Program (ASIP) implement a new version of the SINCGARS radio that (1) is capable of networked IP data communications and (2) adds forward error-correction capabilities, thereby extending the range available for reliable data transmission. ASIP, the software-reprogrammable version of SIP, features reduced size, weight, and power consumption.

Table A.1—Continued

Name	Abbreviation	ACAT	Description
Ships Signal Exploitation Equipment Increment E	SSEE Inc E	III	SSEE Inc E is an evolutionary-development, spiral-acquisition Tactical Cryptologic System whose function is to provide the Navy's surface platforms the capability to carry out comprehensive surface tactical information-warfare exploitation and electronic-warfare support measures.
Ships Signal Exploitation Equipment Increment F	SSEE Inc F	III	SSEE Inc F is an evolutionary-development, spiral-acquisition Tactical Cryptologic System whose function is to provide the Navy's surface platforms the capability to carry out comprehensive surface tactical information-warfare exploitation and electronic-warfare support measures.
Submarine Local Area Network	SubLAN	III	This network provides submarines with mission-essential unclassified and top-secret LANs and a mission-critical secret LAN. When combined with other subsystems, it delivers an end-to-end network-centric–warfare capability. It provides network infrastructure, including a UWLAN, servers, PCs, and COMPOSE.
TacMobile	None	III	The TacMobile program provides fixed-site and mobile C4I warfighting capability to Navy commanders (and their subordinate commands) who support all facets of expeditionary warfare, including littoral, open-ocean, and land operations. Increment 2 includes CMFWIES (non-C2) upgrades for TSC and MOCC. Future upgrades will provide an extensive upgrade to the C2I component, refresh a number of non-C2I components, and supply the new logical interfaces for the P-8A.
Battle Force Email 66 AN/ UYQ-92 (V) 1–4	BFEM 66	IV	BFEM 66 provides an allied-interoperable secure email capability over an HF RF. This system supports a half-duplex, Carrier Sense Multiple Access email net by means of temporary point-to-point links between ship pairs on a single frequency.
Digital Wideband Transmission System	DWTS	IV	DWTS provides up-to-2,048-Kbps ship-to-ship and ship-to-shore data-transmission links. It is employed primarily by ARG/MEU staff for planning and operations. It also provides an interoperable data link with Army and Marine systems.

Table A.1—Continued

Name	Abbreviation	ACAT	Description
Environmental Satellite Receiver Processor Systems	ESRP	IV	The ESRP is a ship/shore system that provides direct download, storage, and processing of raw digital data transmitted from meteorological satellites. The data provide Navy fleet operations with secure, high-resolution, direct-readout imagery, both visual and infrared. This information is used across a broad spectrum of warfare areas, including strike, surface, air, and undersea, and for general weather forecasting. The ESRP receives information from the following satellites: the Defense Meteorological Satellite Program, the Television Infrared Observing System, the Geostationary Operational Environmental Satellite, the Geodetic Satellite Follow-On, the Sea-Viewing Wide Field of View Sensor, the TerraAquaOceanSat (India), the GMS (Japan), the Metosat (European Space Agency), and the Fengyun (China).
Marine Corps Meteorological Mobile Facility (Replacement)	METMF (R)	IV	The METMF(R) is a transportable system that provides tactical METOC to the MAGTF in garrison and while engaged in operations from the sea, sustained operations ashore, and operations other than war. Housed in a single ISO shelter, the METMF(R) can be transported in a single standardized C-130 for rapid deployment operations. The METMF(R) is capable of providing the MAGTF with continuous meteorological observations, satellite imagery, forecasts, and other tactical decision aids and products for 30 days without resupply. The METMF(R) was designed to be interoperable with Marine Corps C4I systems and the meteorology and oceanography systems of the other services and government agencies.
Naval Modular Automated Communication System II/Single Messaging Solution	NAVMACS II/ SMS	IV	NAVMACS II/SMS (also called Tactical Messaging) is transitioning to an inactive program. Legacy NAVMACS systems sustainment and SubSMS procurement will continue until the Assured Internet Protocol is implemented in FY11. Current technology solutions (e.g., DMS Proxy) will be funded and managed within the ISNS ACAT II POR as part of ISNS.

Table A.1—Continued

Name	Abbreviation	ACAT	Description
Navigation Sensor System Interface AN/SSN-6 (V)	NAVSSI	IV	The NAVSSI program's main function is the collection, processing, integration, and distribution of navigation data to weapon systems, combat support systems, C4ISR systems, and other information-system users. These systems depend on NAVSSI to provide critical positioning, navigation, and timing data.
Tactical Switching	TSw	IV	TSw delivers the Navy's shore net-centric infrastructure. Its communication-control nodes will be interoperable with joint, allied, and coalition networks. The program will meet the FORCEnet requirement for net-centric operations by delivering an end-to-end, secure, reliable, reconfigurable, and sustainable C2 network. With an all-IP network as its cornerstone, TSw will help enable the full integration of all Navy voice, video, and data networks into all services within the GIG.
Television Direct to Sailors	TV-DTS	IV	The TV-DTS system is comprised of an OE-556U terminal and an antenna-control unit, both of which interface with the Shipboard Information, Training and Entertainment System for distribution throughout a ship. TV-DTS is a receive-only system that broadcasts three television channels, two music-radio channels, a news and sport–radio channel, and a channel reserved for public-affairs print products.
Advanced Communications Package	ACP	Pre-ACAT	ACP provides over-the-horizon data-relay capability to allow operation of off-board sensors (e.g., ISR, MIW, SUW, USW) at extended ranges.
Advanced High Data Rate Antenna System	AdvHDR	Pre-ACAT	AdvHDR is a next-generation SubHDR multiband antenna for wideband and protected communications. It utilizes phased-array technology and provides two-way wideband capability for K/Ka/Q SATCOM, including EHF MDR, Advanced EHF, Wideband Gap filler, GlobalHawk, JSTARS, and emerging military and commercial satellite systems.

Table A.1—Continued

Name	Abbreviation	ACAT	Description
CENTRIXS-M	CENTRIXS-M	Pre-ACAT	CENTRIXS-M is a Web-centric, COTS-based global network that permits multinational information sharing and provides information services, such as email, Web services, and collaboration, to allied and coalition forces.
Computer Network Defense Increment I	CND	Pre-ACAT	The plan for integrating CND afloat with CANES is currently being determined. Shore requirements may be met through the NGEN and ESSG contracts.
Cross-Domain Solutions Boundary Device	CDS	Pre-ACAT	The overarching CDS capabilities are identified by the GIG, GIG ES, GIG IA, and MNIS ICDs. CDS will be a family of systems that includes CDS Boundary Device, CDS Client Display/Visualization, and Advanced Multi-Level Analysis capabilities. A CDS Boundary Device facilitates data flow between information domains that are normally closed to one another. It includes technology, policy, and threat to ensure that (1) only the intended data reach the destination domain, (2) only the intended data transit the CDS infrastructure, and (3) the intended data always traverse the infrastructure, unless altered by policy or a human user.
Distributed Information Operations– Services	DIO-S	Pre-ACAT	DIO-S will consist of specific services that are available on the network. These DIO services will provide the ability to "telescope" between fine-grain resolution at the tactical level and major trends at the operational and strategic levels. These services need to be scalable in the sense that they must work at multiple levels: the level of an individual platform with a few sensors, the level of an Expeditionary Strike Group consisting of multiple platforms, the level of the fleet, and the regional or global level of the Fleet Information Operations Center. The source and classification level of each piece of data and information need to be tagged at each operational level, and the data and information must then be made available to the network and integrated in the overall database maintained by the Fleet Information Operations Center. At each operational level, access to and feedback to "national" and "tactical" sources are required. Because the DIO-S will be net-centric, these DIO services must be bandwidth-efficient processes.

Table A.1—Continued

Name	Abbreviation	ACAT	Description
Joint Integrated System Technology (CP)	JIST (CP)	Pre-ACAT	This is a congressionally mandated effort to study MILSATCOM planning, management, control capabilities, and associated planning tools. Congress requested that project management be transferred from the Air Force to the Navy to better meet requirements, deadlines, and funding priorities.
Littoral Battlespace Sensing–Unmanned Undersea Vehicles Increment	LBS-UUV	Pre-ACAT	The LBS-UUVs will provide the Joint Force Commander, the Joint Forces Maritime Component Commander, or Combatant Commanders an increased collection capability for oceanographic data in support of Navy antisubmarine warfare, mine warfare, and special-warfare operations. LBS-UUVs improve the coverage, accuracy, and precision of environmental characterizations, enabling the warfighter to make tactical adjustments to asset allocation by optimizing sensor and weapon-platform placement and modes of operation and by increasing tactical effectiveness while reducing tactical timelines and risk to forces. The LBS-UUV capability is comprised of ocean gliders and autonomous undersea vehicles. Its estimated initial operational capability is FY10, and its estimated full operational capability is FY11. Gliders are small (man-portable), long-endurance (weeks to months), buoyancy-driven vehicles that provide a low-cost, semi-autonomous, and highly persistent means of sampling and characterizing the properties of the ocean water column at spatial and temporal resolutions not possible using T-AGS 60 *Pathfinder*-class military survey vessels or tactical units. Autonomous undersea vehicles are larger, shorter-endurance (hours to days), conventionally powered (typically by an electric motor) vehicles that will increase the spatial extent and resolution of the bathymetry data collected by T-AGS–class ships.
Low Band Universal Communications System	LBUCS	Pre-ACAT	This effort will provide an end-to-end open system for a VLF/LF strategic and tactical submarine/shore communications system. This system will replace the legacy AN/URT-30B VLF/LF transmit system, the AN/WRR-12A VLF/LF receive system, and associated cryptographic equipment.

Table A.1—Continued

Name	Abbreviation	ACAT	Description
Naval Integrated Tactical Environmental Support System– Next	NITES-Next	Pre-ACAT	AN/UMK-4(V) NITES is an evolutionary upgrade to the AN/UMK-3(V) Tactical Environmental Support System. NITES-Next will be the latest version of the AN/UMK-4(V) NITES, providing significant improvements over the existing implementation. NITES-Next will provide meteorological and oceanographic support to Navy, Marine Corps, and joint forces engaged in worldwide operations, ashore and afloat. The main functions of the system are to collect, store, and forecast METOC information; assess the impact of present METOC conditions and forecast the impact of future METOC conditions on operations, weapon systems, and sensor systems; and provide METOC data to the warfighter's mission-planning and decision-support systems.
Nuclear Command, Control, and Communications Long Term Solution	NC3 LTS	Pre-ACAT	NC3 LTS will replace NC3 HS and provide critical communications in support of the Joint Operational Architecture for time-critical and non–time-critical communications to be disseminated across areas of operation in support of the new Triad Strategic Communications Emergency Action Message system and data.
Satellite Signals Navigation Set	AN/WRN-X	Pre-ACAT	Replacement/Modernized GPS shipboard receiver provides position, navigation, and timing data for numerous C2 systems aboard ships and is a replacement for the WRN-6 GPS receiver.

Table A.1—Continued

Name	Abbreviation	ACAT	Description
Transformational Communications	TC	Pre-ACAT	The TC program will fulfill the requirements for the Navy Future SATCOM Terminal, which is the next-iteration (i.e., fifth) Navy SATCOM terminal for Navy shore, surface, and subsurface platforms. It will satisfy the Navy's current and future SATCOM requirements. The Future SATCOM Terminal will be developed with evolutionary or revolutionary technologies, and emphasis will be placed on both performance and affordability. A key aspect of the Future SATCOM Terminal will be its use of a single terminal architecture to provide Q-, Ka-, and X-band communications capability using TC, Advanced EHF, Milstar II, Polar and Wideband Global (i.e., X, Ka, and GBS) satellites. The terminal will provide a scalable collection of hardware and software that can be configured to satisfy the communications requirements of individual Navy platforms, providing the net-centric connection to the GIG necessary to support the warfighter in 2017–2025.
Automatic Identification System	AIS	RDC	AIS is a shipboard broadcast-transponder system that is capable of sending and receiving ship information, including GPS (position, course, speed), identification (name, call sign, length, beam), and cargo (draft, type, destination) information.
Commercial Broadband Satellite Program	CBSP	RDC	CBSP is an outgrowth of the Commercial Wideband Satellite Program and Inmarsat communications. CBSP will provide additional commercial SATCOM throughput to support emergent fleet operational requirements and to counter potential enemy threats while providing full redundancy in the event that MILSATCOM becomes unavailable.
Expanded Maritime Intercept Operations	EMIO	RDC	The EMIO wireless reach-back initiative provides a secure wireless-transmission system capable of transmitting EMIO-specific data from visit, board, search, and seizure teams aboard the target vessel to the on-scene command ship.
Maritime Domain Awareness Prototype Acceleration	MDA	RDC	The MDA Prototype Acceleration program responds to a May 17, 2007, memo from the Secretary of the Navy calling for accelerated delivery of MDA capabilities to the fleet.

Table A.1—Continued

Name	Abbreviation	ACAT	Description
Subnet Relay and High Frequency Internet Protocol	SNR/HFIP	RDC	SNR is a line-of-sight UHF capability that provides a multinode, multihop (up to four hops), ship-to-ship network using IP protocols. SNR will improve interoperability, reduce reliance on SATCOM, and provide communications redundancy.

SOURCE: Program Executive Office, Command, Control, Communications, Computer and Intelligence, unnamed Web page, undated-b.

Descriptions of Analyzed Upgrades

Table B.1 supplies amplifying information—including the program manager, alternative identifiers, and descriptions—of the upgrades analyzed during our study.

Table B.1
Descriptions of Analyzed Upgrades

Program Manager	System Name	Alternative Identifier	Alternative Brief Identifier	Upgrade Description
PMW 120	SSEE Increment E	SA CG 0047 00734 K 00	SSEE INC E Upgrade	Replaces the existing SSEE system within SSES with four new, state-of-the-art COTS equipment racks. Also reconfigures the legacy SSES communications equipment located in six CCSS racks into a suite of four racks (including NOW terminals).
PMW 120	SSEE Increment E	SA CVN 0068 73773 K 00	SSEE INC E VER 4.0	Replaces the existing BGPHES system within SSES with four new, state-of-the-art COTS equipment racks. Also replaces the existing AS4293 antennas with new or refurbished antennas of the same fit, form, and function. Replaces forward HF Whip antennas (currently, AS2537As with the Antenna Tilt Group) with two AS-142s. Also removes CDLS.
PMW 120	SSEE Increment E	SA DDG 0051 71377 K 00	SSEE INC E SHIPALT	Replaces the existing 12-rack Combat DF System with five new, state-of-the-art COTS equipment racks within the DDG51FLT II/IIA-Class designated SSES space. Also installs two OE-578/USQ antennas and replaces two existing AS-2867/SRR antennas with upgraded AS2867B/SRR models. One existing AS3606/IUC-109 antenna will be replaced with a new AS142 acquisition antenna.
PMW 120	SSEE Increment E	SA LHD 0001 00448 K 00	SSEE INC E Upgrade	Replaces the existing Combat DF System (AN/SRS-1(V)7) on LHDs 1–4 with seven new, state-of-the-art COTS equipment racks within the designated SSES space. Also replaces the OE-326(V)1 antenna with an upgraded OE-326(V)5.

Table B.1—Continued

Program Manager	System Name	Alternative Identifier	Alternative Brief Identifier	Upgrade Description
PMW 120	SSEE Increment E	SA LHD 0001 03264 K 00	SSEE INC E Upgrade	Replaces the existing Combat DF System (AN/SRS-1(V)8) on LHDs 5–8 with seven new, state-of-the-art COTS equipment racks within the designated SSES space. Also replaces the OE-326(V)1 antenna with an upgraded OE-326(V)5.
PMW 120	SSEE Increment E	SA LPD 0017 71163 K 00	SSEE INC E	Installs four new, state-of-the-art COTS equipment racks within SSES. Also installs rack foundations and sway bracing, reroutes HVAC ducting, installs platforms and cabling to accommodate two OE-578 antennas, and installs foundations and cabling for HF coaxial interface equipment within the radio-transmitter room.
PMW 150	GCCS-M 4.X	SA CG 0047 00733 K 00	GCCS-M GENSER 4.X (V) 4	Removes all existing GCCS-M HP UNIX servers and workstations on the GENSER ISNS network and upgrades the existing rack with an open-ended architecture system (within the rack). Also installs new shock coils and dual-stage kits. Replaces three existing HP UNIX workstations (in the CIC) and four PC workstations with seven Windows 2000 PC clients.

Table B.1—Continued

Program Manager	System Name	Alternative Identifier	Alternative Brief Identifier	Upgrade Description
PMW 150	GCCS-M 4.X	SA CVN 0071 09309 K 00	GCCS-M GENSER 4.X (V)2	Provides for the technology refresh of legacy HP-based GCCS-M 3.X server racks to the Solaris-based implementation of GCCS-M 4.X. Removes up to six equipment racks and replaces them with two racks. Installs shock isolators and new cooling fans on all server racks. Replaces all existing GCCS-M HP UNIX workstations (up to 60) with the applicable COMPOSE PCs and peripherals. Also installs two new Solaris GALE Lite workstations. Installs GCCS-M 4.X software.
PMW 150	GCCS-M 4.X	SA CVN 0071 09310 K 00	GCCS-M SCI 4.X (V)2	Removes all existing GCCS-M HP UNIX servers and workstations on the SCI network. Installs five Solaris UNIX servers in two existing SCI racks. Installs three PC servers in existing server racks. Replaces all existing GCCS-M HP UNIX workstations with Windows 2000 PCs. Installs two new Solaris GALE Lite UNIX workstations. Installs GCCS-M 4.X software.
PMW 150	GCCS-M 4.X	SA CVN 0073 09310 K 00	GCCS-M SCI 4.X (V)2	Removes all existing GCCS-M HP UNIX servers and workstations on the SCI network. Installs five Solaris UNIX servers in two existing SCI racks. Installs three PC servers in existing server racks. Replaces all existing GCCS-M HP UNIX workstations with Windows 2000 PCs. Installs two new Solaris GALE Lite UNIX workstations. Installs GCCS-M 4.X software.

Table B.1—Continued

Program Manager	System Name	Alternative Identifier	Alternative Brief Identifier	Upgrade Description
PMW 150	GCCS-M 4.X	SA CVN 0074 09310 K 00	GCCS-M SCI 4.X (V)2	Removes all existing GCCS-M HP UNIX servers and workstations on the SCI network. Installs five Solarix UNIX servers in two existing SCI racks. Installs three PC servers in existing server racks. Replaces all existing GCCS-M HP UNIX workstations with Windows 2000 PCs. Installs two new Solaris GALE Lite UNIX workstations. Installs GCCS-M 4.X software.
PMW 150	GCCS-M 4.X	SA DDG 0051 00433 K 00	GCCS-M GENSER 4.X (V)4	Removes all existing GCCS-M HP UNIX servers and workstations on the GENSER ISNS network and upgrades the existing rack with an open-ended architecture system (within the rack). Also installs new shock coils and dual-stage kits. Replaces three existing HP UNIX workstations (in the CIC) and installs a database-manager rack with three client computers. Also replaces four existing remote SIPRNET PC workstations with four new GCCS-M/SIPRNET PC workstations.
PMW 150	GCCS-M 4.X	SA LHD 0001 00427 K 00	GCCS-M GENSER 4.X (V)3	Provides for the technology refresh of legacy HP-based GCCS-M 3.X server racks to the Solaris-based implementation of GCCS-M 4.X. Removes up to four equipment racks and replaces them with two racks. Installs shock isolators and new cooling fans on all server racks. Replaces all existing GCCS-M HP UNIX workstations (up to 60) with the applicable COMPOSE PCs and peripherals. Also installs two new Solaris GALE Lite workstations. Installs GCCS-M 4.X software.

Table B.1—Continued

Program Manager	System Name	Alternative Identifier	Alternative Brief Identifier	Upgrade Description
PMW 150	GCCS-M 4.X	SA LHD 0001 00429 K 00	GCCS-M SCI 4.X (V)3	Removes all existing GCCS-M HP UNIX servers and workstations on the SCI network. Installs five Solarix UNIX servers in two existing SCI racks. Installs three PC servers in existing server racks. Replaces all existing GCCS-M HP UNIX workstations with Windows 2000 PCs. Installs two new Solaris GALE Lite UNIX workstations. Installs GCCS-M 4.X software.
PMW 150	GCCS-M 4.X	SA LPD 0004 70431 K 00	GCCS-M 4.0 USQ-172(V)5	Provides for the technology refresh of legacy HP-based GCCS-M 3.X server racks to the Solaris-based implementation of GCCS-M 4.X. Upgrades one server rack in the CIC. Installs shock isolators and new cooling fans on all server racks. Replaces all existing GCCS-M HP laptop computers with COMPOSE laptops. Installs GCCS-M 4.X software.
PMW 150	GCCS-M 4.X	SA SSN 0688 04323 P 03	Install GCCS-M 4.X	Installs lockers inside of the small rack on the first platform, but only if 4.01 and 4.03 are installed together without 4.02. Enables the 4.X Web upgrade of the TIDS 1 network.
PMW 160	ISNS	SA CG 0047 00654 K 00	Embarkable Drops	Installs additional ISNS classified/unclassified PC and printer drops by installing additional 10/100TX or 100FX network modules into existing shipboard networking switches and installing associated STP cabling and connection boxes in the desktop. May require the use of an OmniStack in the area.

Table B.1—Continued

Program Manager	System Name	Alternative Identifier	Alternative Brief Identifier	Upgrade Description
PMW 160	ISNS	SA CVN 0073 08922 K 00	Embarkable Drops	Installs additional ISNS classified/unclassified PC and printer drops by installing additional 10/100TX or 100FX network modules into existing shipboard networking switches and installing associated STP cabling and connection boxes in the desktop. May require the use of an OmniStack in the area.
PMW 160	ISNS	SA DDG 0051 00378 K 00	Embarkable Drops	Installs additional ISNS classified/unclassified PC and printer drops by installing additional 10/100TX or 100FX network modules into existing shipboard networking switches and installing associated STP cabling and connection boxes in the desktop. May require the use of an OmniStack in the area.
PMW 160	ISNS	SA FFG 0007 00444 K 00	Embarkable Drops	Installs additional ISNS classified/unclassified PC and printer drops by installing additional 10/100TX or 100FX network modules into existing shipboard networking switches and installing associated STP cabling and connection boxes in the desktop. May require the use of an OmniStack in the area.

Table B.1—Continued

Program Manager	System Name	Alternative Identifier	Alternative Brief Identifier	Upgrade Description
PMW 160	ISNS	SA LSD 0041 01315 K 00	Embarkable Drops	Installs additional ISNS classified/unclassified PC and printer drops by installing additional 10/100TX network modules into existing shipboard networking switches and installing associated UTP (or STP for classified) cabling and connection boxes in the desktop. May require the use of an OmniStack in the area.
PMW 160	ISNS	SA CG 0047 74058 K 00	ISNS AN/USQ-153 Aegis AWS/ADS	Installs the necessary Fiber Distributed Data Interface modules within the two ISNS SECRET Backbone Switch Racks. Includes fiber-optic cable for all the required connections to support an SA CG 47 489K connection to the ADS router with the ISNS SECRET Backbone switches.
PMW 160	ISNS	SA CG 0047 00714 K 00	ISNS LAN GIG-E	Installs the host ISNS backbone LAN and associated interconnections to other systems, drops, associated hardware (i.e., fiber-optic cable, servers, routers, and switches), and operating software necessary for initial installation. If a ship has received an ISNS ATM LAN, the current switches with enclosures and server suites are replaced with new equipment in racks. In all other cases, any old equipment, including enclosures, is removed, and a completely new LAN, including racks and drops, is installed.

Table B.1—Continued

Program Manager	System Name	Alternative Identifier	Alternative Brief Identifier	Upgrade Description
PMW 160	ISNS	SA CVN 0073 09323 K 00	ISNS LAN GIG-E	Installs the host ISNS backbone LAN and associated interconnections to other systems, drops, associated hardware (i.e., fiber-optic cable, servers, routers, and switches), and operating software necessary for initial installation. If a ship has received an ISNS ATM LAN, the current switches with enclosures and server suites are replaced with new equipment in racks. In all other cases, any old equipment, including enclosures, is removed, and a completely new LAN, including racks and drops, is installed.
PMW 160	ISNS	SA DDG 0051 00425 K 00	ISNS LAN GIG-E	Installs the host ISNS backbone LAN and associated interconnections to other systems, drops, associated hardware (i.e., fiber-optic cable, servers, routers, and switches), and operating software necessary for initial installation. If a ship has received an ISNS ATM LAN, the current switches with enclosures and server suites are replaced with new equipment in racks. In all other cases, any old equipment, including enclosures, is removed, and a completely new LAN, including racks and drops, is installed.

Table B.1—Continued

Program Manager	System Name	Alternative Identifier	Alternative Brief Identifier	Upgrade Description
PMW 160	ISNS	SA FFG 0007 00467 K 00	ISNS LAN GIG-E	Installs the host ISNS backbone LAN and associated interconnections to other systems, drops, associated hardware (i.e., fiber-optic cable, servers, routers, and switches), and operating software necessary for initial installation. If a ship has received an ISNS ATM LAN, the current switches with enclosures and server suites are replaced with new equipment in racks. In all other cases, any old equipment, including enclosures, is removed, and a completely new LAN, including racks and drops, is installed.
PMW 160	ISNS	SA LSD 0041 01332 K 00	ISNS LAN GIG-E	Installs the host ISNS backbone LAN and associated interconnections to other systems, drops, associated hardware (i.e., fiber-optic cable, servers, routers, and switches), and operating software necessary for initial installation. If a ship has received an ISNS ATM LAN, the current switches with enclosures and server suites are replaced with new equipment in racks. In all other cases, any old equipment, including enclosures, is removed, and a completely new LAN, including racks and drops, is installed.

Table B.1—Continued

Program Manager	System Name	Alternative Identifier	Alternative Brief Identifier	Upgrade Description
PMW 160	ISNS	SA MCM 0001 70475 K 00	LAN ISNS GIG-E (AN/USQ-153 V8)	Installs a new GIG-E LAN, which consists of two equipment racks (unclassified and secret enclaves), each of which contains a router, switches, servers, and UPS and associated interconnections to other systems and drops. CFCP and JMCIS 98 components are removed. ISNS PCs are added to existing ISNS PCs so that the final ISNS PC quantity totals to the amount funded in the PC Refresh Plan. PCs that are already on board will be either replaced (if obsolete) or left intact (if not obsolete). Classified PCs may be laptops rather than desktops.
PMW 160	ISNS	EC 71021 43 AN/USQ-153(V)	ISNS COMPOSE 3.0 SW Install	Installs COMPOSE v3.0. The ISNS ECO 43 COMPOSE 3.0 Software Load will be used to implement this release. Depending on the ship's ISNS variant configuration, an additional server or servers may be required to support the COMPOSE v3.0 load.

Table B.1—Continued

Program Manager	System Name	Alternative Identifier	Alternative Brief Identifier	Upgrade Description
PMW 160	ISNS	EC 73836 ISNS ECO 62 ATM SLEP	ISNS ECO 62 ATM SLEP	Installs the upgrade to the ISNS AN/USQ-153(V)1 LAN in an effort to provide increased data flow within the existing architecture to support operations prior to the installation of the AN/USQ-153A(V)1 GIG-E LAN. Removes MSS from the setup as the OC-3 link from the backbone switch to the MSS. Provides MPM-III modules in place of the MPM-1Gs in all of the switches. For the LCC-20, LHD-3 upgrades all the FCSMs to FCSM-IIs for improved performance of the conversion of data from Ethernet to ATM, and vice versa. The Alcatel Operating System (XOS) is upgraded to the latest available version (4.4.2) supported by the newer MPM-III modules.
PMW 160	ISNS	EC 73839 ISNS LAN SLEP & GIGE LANS	ISNS LAN SLEP & GIGE EC63	Installs the upgrade to the ISNS AN/USQ-153(V)1 LAN in an effort to provide increased data flow within the existing architecture to support operations prior to the installation of the AN/USQ-153A(V)1 GIG-E LAN. This is accomplished by providing upgraded MPMs in the LAN edge switches. To provide increased data flow, MPM-III modules replace the MPM-1Gs provided with EC17R3 (the GIG-E Backfit) in all of the switches. An operating-system upgrade is installed in the backbone switches (if this has not already occurred). The Alcatel Operating System (XOS) is upgraded to the latest available version (4.4.2) supported by the newer MPM-III modules.

Table B.1—Continued

Program Manager	System Name	Alternative Identifier	Alternative Brief Identifier	Upgrade Description
PMW 170	SHF	SA CG 0047 00773 K 00	EBEM for WSC-6 Variants	Installs the upgrade to the modem assembly of the AN/WSC-6A(V)7 SHF SATCOM with the MD-1366A/U EBEM. Installs one MD-1366A/U EBEM, based on the system version, external to the SHF equipment rack. Installation includes interconnecting cables.
PMW 170	SHF	SA DDG 0051 00471 K 00	EBEM for WSC-6 Variants	Installs the upgrade to the modem assembly of the AN/WSC-6A(V)9 SHF SATCOM with the MD-1366A/U EBEM. Installs two MD-1366A/U EBEMs external to the SHF equipment racks. Installation includes interconnecting cables.
PMW 170	SHF	SA LCC 0019 01543 K 00	EBEM for WSC-6 Variants	Installs the upgrade to the modem assembly of the AN/WSC-6A(V)5 SHF SATCOM with the MD-1366A/U EBEM. Removes the existing, obsolete CQM-248A modems and/or SLM-3650 modems and installs two MD-1366A/U EBEMs external to the SHF equipment racks.
PMW 170	SHF	SA LPD 0004 01342 K 00	EBEM for WSC-6 Variants	Installs the upgrade to the modem assembly of the AN/WSC-6A(V)7 SHF SATCOM with the MD-1366A/U EBEM. Installs one MD-1366A/U EBEM, based on the system version, external to the SHF equipment rack. Installation includes interconnecting cables.

Table B.1—Continued

Program Manager	System Name	Alternative Identifier	Alternative Brief Identifier	Upgrade Description
PMW 170	SHF	SA DDG 0051 71556 K 00	SHF (AN/WSC-6E(V)9)	Installs the AN/WSC-6E(V)9 on DDG-51-class ships. This alteration installs the AN/WSC-6E(V)9 System CY-8893/WSC-6E(V)9 in the equipment cabinet, baseband equipment in the communications center, and two OE-580/WSC-6(V) antenna pedestals topside. The antenna installation requires the relocation of the UHF SATCOM antennas, associated amplifier filter boxes, and associated cabling. Ancillary support equipment and services include a waveguide and antenna pedestal dry-air compressor piping system with associated valves and flow meters, ship's conditioned air for cooling (with heating capability directly vented and exhausted into the antenna radomes), power and countermeasures piping topside, and redistribution of communications-center cooling and exhaust ductwork. Baseband equipment occupies an additional 19-inch rack. On ships without EHF FOT, the installation of AN/WSC-6E(V)9 requires relocation of the forward mast light. On ships with existing dual Inmarsat systems, the second Inmarsat system is removed upon installation of the AN/WSC-6E(V)9.

Bibliography

Adan, Joe, "Creating the Global Maritime Security Force," briefing, AFCEA 2007 Joint C4ISR Symposium, May 24, 2007.

Defense Information Systems Agency, "Net-Centric Enterprise Services (NCES)," Web page, undated. As of July 28, 2009:
http://www.disa.mil/nces/

Department of Defense, *Mandatory Procedures for Major Defense Acquisition Programs (MDAPs) and Major Automated Information System (MAIS) Programs*, DoD Regulation 5000.2-R, March 15, 1996.

———, *Department of Defense Dictionary of Military and Associated Terms*, Joint Publication 1-02, April 12, 2001, as amended through May 30, 2008.

Department of Defense, Chief Information Officer, *Department of Defense Global Information Grid Architectural Vision: Vision for a Net-Centric, Service-Oriented DoD Enterprise*, version 1.0, June 2007.

Department of the Navy, *Sea Power for a New Era: A Program Guide to the U.S. Navy*, Web page, undated. As of July 28, 2009:
http://www.navy.mil/navydata/policy/seapower/spne06/top-spne06.html

Department of Navy, Research, Development & Acquisition, "GCCS-M," Web page, undated. As of July 28, 2009:
http://acquisition.navy.mil/content/view/full/4695

Deputy Chief of Naval Operations, Warfare Requirements and Programs (N6/N7), *FORCEnet Requirements/Capabilities and Compliance Policy*, Ser N6N7/5U91622 2, May 27, 2005a.

———, *Requirement for Open Architecture (OA) Implementation*, Ser N6N7/5U916276, December 23, 2005b.

Greene, William H., *Econometric Analysis, Fifth Edition*, Upper Saddle River, N.J.: Prentice Hall, 2003.

Iacovetta, Joseph, Jim Balon, Joe Bracewell, Chuck Auxter, Jim Burgess, Mike Cullison, Bernie Dombrosky, Timothy Hickey, Carol Smith, Travis Tillman,

Michael Virnig, and Glen Hoffman, *21st Century C4ISR/IW Technology Insertion and Management*, report presented at the Engineering Total Ship (ETS) 2000 Symposium, Gaithersberg, Md., March 21, 2000. As of July 28, 2009:
http://stinet.dtic.mil/cgi-bin/GetTRDoc?AD=ADA377328&Location=U2&doc=GetTRDoc.pdf

Lee, David A., *The Cost Analyst's Companion*, McLean, Va.: Logistics Management Institute, 1997.

National Research Council, *The Role of Experimentation in Building Future Naval Forces*, The National Academies Press, Washington, D.C., 2004.

————, *C4ISR for Future Naval Strike Groups*, The National Academies Press, Washington, D.C., 2006.

Naval Network Warfare Command, "FORCEnet," Web page, undated. As of July 28, 2009:
http://forcenet.navy.mil/

NAVSEA Shipbuilding Support Office, "Naval Vessel Register: The Official Inventory of US Naval Ships and Service Craft," Web page, date not available. As of July 28, 2009:
http://www.nvr.navy.mil

Office of the Under Secretary of Defense (Comptroller), *National Defense Budget Estimates for FY 2009*, March 2008. As of July 28, 2009:
http://www.defenselink.mil/comptroller/defbudget/fy2009/fy2009_greenbook.pdf

PEO C4I—*see* Program Executive Office, Command, Control, Communications, Computers, and Intelligence.

Program Executive Office, Command, Control, Communications, Computers, and Intelligence, "Information Dominance; Anytime, Anywhere . . . ," Web page, undated-a. As of July 28, 2009:
http://enterprise.spawar.navy.mil/body.cfm?type=c&category=38&subcat=180

————, unnamed Web page, undated-b. As of July 28, 2009:
http://enterprise.spawar.navy.mil/body.cfm?type=c&category=38&subcat=196

————, "Program Executive Office, Command, Control, Communications, Computers, and Intelligence (C4I)," fact sheet, March 14, 2008. As of July 28, 2009:
http://enterprise.spawar.navy.mil/cmt_uploads/38/FactSheet_031408-S.pdf

Program Executive Office, Command, Control, Communications, Computers, and Intelligence (PEO C4I)/Networks, Information Assurance and Enterprise Services Program Office (PMW 160), "Networks, IA and Enterprise Services, PMW 160 Networks Overview, Veteran Small Business Conference," briefing, San Diego, Calif., June 2007.

Secretary of the Navy, *Implementation of Mandatory Procedures for Major and Non-Major Defense Acquisition Programs and Major and Non-Major Information Technology Acquisition Programs*, SECNAVINST 5000.2B, December 6, 1996.

Space and Naval Warfare Systems Command, *21st Century C4ISR/IW Technology Insertion and Management*, Dahlgren, Va., Naval Surface Warfare Center, 2000.

Turner, Phil, "The CANES Initiative: Bringing the Navy Warfighter onto the Global Information Grid," *CHIPS*, October/December 2007, pp. 24–27.

U.S. Navy Program Executive Officer, Command, Control, Communications, Computers, and Intelligence Team, *Navy C4I Open Architecture Strategy*, San Diego, Calif., 2007a.

—————, *Strategic Plan*, San Diego, Calif., 2007b.

Walsh, Ed, *Aegis Aims for Open Architectures by 2007*, U.S. Naval Institute Proceedings, February 2005.

Witten, Ian H., and Eibe Frank, *Data Mining: Practical Machine Learning Tools and Techniques*, Oxford, UK: Elsevier, 2005.

Yardley, Roland J., James G. Kallimani, John F. Schank, and Clifford A. Grammich, *Increasing Aircraft Carrier Forward Presence: Changing the Length of the Maintenance Cycle*, Santa Monica, Calif.: RAND Corporation, MG-706-NAVY, 2008. As of July 7, 2009:
http://www.rand.org/pubs/monographs/MG706/